D0833696

ENGLAND'S FLORA

EDWARD HYAMS

THE STORY OF
ENGLAND'S FLORA

Illustrated by Josephine Ranken

BOOK CLUB ASSOCIATES

LONDON

This edition published 1979 by Book Club Associates
By arrangement with Kestrel Books.

First published 1979

Printed in Great Britain by Morrison and Gibb Ltd
London and Edinburgh

Contents

For John Guest, knowing he will enjoy it

Publisher's Note

Sadly, Edward Hyams died before he could see this book through the press. He had been involved in the early stages of planning the layout of the book; he had corrected galleys of the text; and he had seen and approved draft versions of the illustrations. We hope therefore that the finished book is very much what he had in mind when he set out to describe the story of England's flora, and that it will act as a fitting memorial to him.

The publishers would like to thank Mr Anthony Huxley for his advisory work on the book – both text and illustrations – before it went to press.

The Layout of the Book

Chapters 2–5 form the core of the book in so far as they discuss the history of a great number of plants and trees found today in England. Whenever a particular plant is under discussion in these chapters, the first use of its name is given in bold for convenience. Each of these chapters ends with a list of all the species mentioned in the chapter, giving the correct botanical name and the family to which the plant belongs.

Appreciation

Edward Hyams was an authority and writer on many subjects, but they had a basic common theme: as one of his friends wrote after his death, his special quality 'lay in his ability to become interested in absolutely everything that concerned the life and civilization of man'.

One of his greatest interests was gardens, their design and history, about which he wrote several major books including *The English Garden, The Irish garden, A History of Gardens and Gardening* and *Great Botanical Gardens of the World*. He had practised vine growing, and wrote several books on viticulture and wine, including an entertaining social history called *Dionysus*. He was specially interested in the plants man makes use of for food, spice, drink and medicine – a subject closely linked with those of archaeology and how man lived in the past – which formed the theme of *Plants in the Service of Man*. In the 1950s he produced two of the earliest books on the vexed questions of conservation – *From the Waste Land* and *Soil and Civilization*.

As well as all these basically factual books he wrote a number of novels and short stories: he was certainly a man of many parts. Not least, he was a practical man: he made several interesting gardens for himself and, on a larger scale, acted as consultant to the Government of Iran for the design and construction of the new Botanic Garden outside Tehran.

Few other people could have produced this fascinating book on the origins, development and uses of the plants we are likely to see around us. The last he wrote, it reflects and links together many of his past specialities.

ANTHONY HUXLEY

September 1978

Introduction

IN this book I shall try to give some idea of what the English flora was like before man had done anything to change it; how it got to be as it was; and then to sketch what happened during the six or seven thousand years of man's history in England to change the wild and cultivated flora into what it is today. England began with a relatively poor flora, by comparison with, say, France or Spain; if we take the cultivated part of it into account, she now has one of the richest in Europe and for the most part that is the work of man. Even strictly wild flora is at the very least 25 per cent richer in species than it was before man began to change it.

As far as we know, all life began in the sea, vegetable as well as animal. That being so, we can safely say that the first plants were water plants, that the sea, lakes and rivers had a rich flora long before there was a single plant growing on land and indeed before there was anything we should recognize as suitable conditions for plants to grow in. The most ancient genera of water plants, hundreds of millions of years old, are still with us, or descendant species very like them. For example, the algae began as microscopic unicellular plants; and although from some such beginning all plant life evolved, there is still an immense population of just such minute and simple plants: they are the green slime which, in certain conditions, accumulates in condensation on greenhouse windows.

Not all of the algae are unicellular micro-organisms; some are very large and complicated-looking seaweeds with apparent 'stems' and apparent 'leaves' – though they are neither in the proper sense of those words. But further back in time and evolution than the unicellular algae we cannot go.

These sea-water and freshwater plants, although simpler in structure than the sort of plants we shall be dealing with, have a more complex life-cycle. And something very like that life-cycle, and the seaweeds' means of reproducing their kind and so propagating their species, is also found in a whole, and very large, group of dry-land plants, the cryptogams – non-flowering plants such as ferns, mosses, liverworts, lichens and fungi. It is obvious that, evolving from seaweeds and water-weeds, the cryptogams are the most ancient of all land-plants, enormously older than the higher plants, gymnosperms and angiosperms,*

*Gymnosperms are those higher plants whose seeds are not enclosed in an ovary – the conifers, for example. Angiosperms are the flowering plants which bear their seeds in an ovary, that is the vast majority of plants in our flora.

which this book is about. The most fundamental difference between the cryptogams and the flowering plants is this: the cryptogams reproduce themselves by releasing astronomical numbers of microscopic spores which contain no plant embryo and which, following germination, turn into embryo plants; whereas all the others, the flowering plants, reproduce themselves by bearing seed in which there is already, no matter how tiny the seed, an embryo plant.

So we can begin by saying that the most ancient vegetation of what is now England consisted of the ferns and mosses and lichens and liverworts. About fifty species of survivors from that unimaginably remote age are still to be found growing wild in England today; when you see a fern, salute it; it is of a lineage at least five hundred times as old as yours.

Our own flora is predominantly one of the flowering and seed-bearing plants so that what we want to know, if we are to complete the chain of links connecting it to the unicellular algae, is – when did the first of such plants evolve? It is an unanswerable question because a lot of links are missing. Still, we do know a little about it: for example, an organ peculiar to the flowering, seed-bearing plants is the pollen grain; the cryptogams have nothing like it.

Fossils of pollen-grains similar in structure to those of modern higher plants have been found, in Sweden and in Scotland, in strata at least one hundred and sixty million years old. Perhaps a series of mutations in some genus belonging to the cryptogams gave rise to the first of the flowering plants about two hundred million years ago. Or it may be that the first flowering, seed-bearing plants evolved from a much simpler and earlier evolutionary stage in the history of plants and that we have not yet found fossils of those ancestral angiosperms. In a letter, written in 1881 to Sir Joseph Hooker, the great botanist who made Kew the best botanical garden in the world, Charles Darwin wrote: '*Nothing is more extraordinary in the history of the vegetable kingdom, as it seems to me, than the apparently very sudden or abrupt development of the higher plants.*' What Darwin was referring to was this: in rock formations of the Cretaceous age – over one hundred million years old – were found fossils of flowering plants of species which are still going strong (for example, oaks, planes, poplars and willows); to have reached that stage by a hundred million years ago these plants must surely have been evolved for almost as long again, and the fact that we have found no material evidence for that evolution means only that for some reason it no longer exists or that we have not looked in the right places – now, perhaps, at the bottom of the sea.

Although, then, higher plants, gymnosperms and angiosperms, were in existence, not less than one hundred and fifty million years ago, they seem, from the fossil evidence, to have remained in a subordinate place for many more tens of millions of years, subordinate to the population of cryptogams, whose most majestic representatives were the tree ferns; their doubtless degenerate descendants are still the glory of parts of Australia and New Zealand, historically the newest but geologically the most ancient lands in the world. Not until the late Cretaceous period, which ended sixty million years ago, did flowering plants gain the

Goat willow, *Salix caprea*

Lombardy poplar, *Populus nigra* var. *italica*

10

ascendancy over the ferns and mosses. Why did they do so? I would guess that it was because, by having their 'young' pass the most vulnerable stage, the embryo stage, in a seed instead of requiring the embryo to develop unprotected at large in the world and in ferocious competition, they slowly gained the upper hand. And also that on land plants whose germination required only a small amount of moisture had an advantage over plants whose spores can only germinate in water.

The Tertiary age or epoch is divided into periods, and the oldest period is called the Palaeocene; the next period, the Eocene, lasted sixteen million years from about fifty-four to about thirty-eight million years ago, the later date being at least thirty million years before man's remotest mammalian ancestor had put in an appearance. The geological deposit called London Clay, in the Thames Valley, belongs to that era, and in that rather unpleasant substance have been found the fossil remains of a considerable number of plants. Nearly all of them belong to tropical genera (for example, palms of a kind now native to South-East Asia or Central America) but about 10 per cent of the total are fossils of temperate-zone species, like the giant redwoods (genus *Sequoia* or perhaps *Metasequoia*) and the relatives of the maidenhair tree, *Ginkgo biloba*.

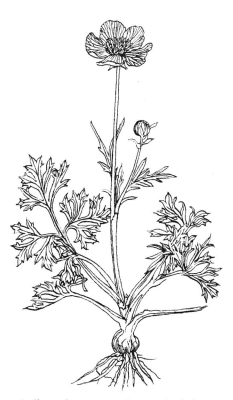

Bulbous buttercup, *Ranunculus bulbosus*

Meanwhile the temperate-zone flora of today had been evolving in what were then temperate but are now subarctic and arctic regions – species of trees, shrubs and herbaceous plants which are not identical with those we know now (with some exceptions), but which we should immediately recognize and have no trouble in identifying. So the generic types of modern plants were already, more than thirty-eight million years ago, well established.

The climate of the region which is now England was meanwhile very slowly changing from tropical through subtropical to temperate; while the tropical flora long persisted in the south, no doubt, in the north it was giving way to a hardier flora migrating southwards from subarctic regions as those regions became colder. So, in the next Tertiary period, which is called Oligocene and which lasted from about thirty-eight million to about twenty-six million years ago, we find, in, for example, the flora of the Isle of Wight, such plants as sedges, poppies, buttercups, oaks, beeches and hornbeams. It is obvious that by the end of the Oligocene the climate of all England had become colder; if the flora is anything to go by, it would seem to have been by then very much what it is today.

The succeeding period is called Miocene and it lasted until about seven million years ago, and during that period the temperate-zone flora migrating south became much richer in England; that is a surmise, for there is very little fossil evidence for that period from English sources; but there *is* fossil evidence from our latitude in Germany and something like five hundred species of temperate-zone tree, shrub and herbaceous species, all of generic types easily recognizable in modern representatives, have been found there. It is very probable that England had much the same flora, though perhaps it was not identical.

The period from about seven million years ago up to only two million years ago, the most recent period of the Tertiary, is called the Pliocene.

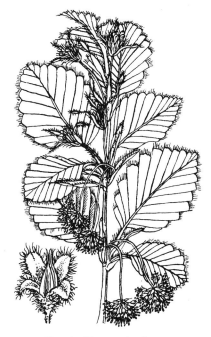

Beech, *Fagus sylvatica*

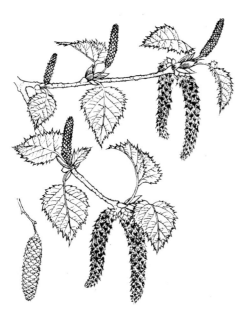

Silver birch, *Betula pendula*

Hornbeam, *Carpinus betulus*

In the course of it the whole northern hemisphere was still growing slowly colder. The process may not have been steady and continuous, but this period certainly ended much colder than it began.

About two million years ago the Ice Age – the Pleistocene – began. This was the beginning of the Quaternary period. The polar ice field began to expand. Consequently an increasing number of the plant species which characterized the earlier period was slowly eliminated from our flora as conditions became less and less congenial. This does not mean that they were exterminated; it means that, as the northern cold increased, they found congenial conditions further south than before. Of course, the climate in the southern part of the northern hemisphere was also changing; what was to be the Ice Age in the more northerly parts was further south the Pluvial Age, a long period of very heavy and continuous rainfall and cool weather in, for example, the lower, but not equatorial, latitudes of North Africa. At no time did the ice reach so far south as to exterminate the tropical flora; and the temperate and subtropical flora survived as far north as the Canary Islands. Parts of southern England, though bitterly cold, were not actually covered by glaciers.

So, before the really bitter cold of the Ice Age reached England, the flora was, as far as we can tell from the fossil record, not very different from that of today, though perhaps not so rich: a few examples from the Cromer Forest Bed geological deposit of the late Pliocene will show what I mean. In that deposit have been found fossils of alder and birch, maple, hornbeam, yew, oak, pine, spruce, elm and beech, in some cases represented by species no longer with us but very close to those which are, and in others (for example, the maple, *Acer campestre*) identical with our familiar species. Among the herbaceous plants are species of Rosaceae, to which the roses and all the fruit trees belong but which also has many herbaceous genera; Compositae, the great family of the daisy genera; Labiatae, that admirable family to which belong the genera of our aromatics such as thyme and rosemary, lavender, sage and marjoram; Umbelliferae, the family which includes such majestic plants as hemlock, angelica and fennel; and Ranunculaceae, the buttercups in all their great variety. What is very curious is that also among these Cromer Forest Bed fossil plants are to be found many of the species we think of as weeds of cultivation (for example, bindweed, chickweed, dock and couchgrass) all ready to annoy the gardener, over a million years ahead of his time.

Chapter One

What We Were Given

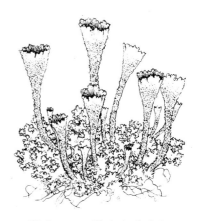

Pixie cups, *Cladonia fimbriata*

THE glaciation of England was a winter lasting nearly two million years. A great part of the temperate-zone flora was wiped out, for even where the ice did not extend the cold was bitter; only the toughest grasses, sub-shrubs and lichens could survive – the arctic flora in fact. But south of England there were places (for example, in North Africa, parts of west Asia, north India, and such enclaves of warmth as the Canary Islands) where this was not so.

Approximately sixteen thousand years ago the glaciers began to contract. During the summers the glaciers shrank by the melting of their ice; with the onset of winter there was more freezing and more snow to become impacted into ice. But, for the first time for thousands of years, the winter freeze could not keep up with, much less exceed, the summer thaw. So there was a net loss of ice year after year. It was not a steady retreat: there were interludes, some of them lasting several centuries, during which the summer thaw was balanced by winter freeze and the retreat of the glaciers was halted. But each time it was eventually resumed until, at last, the ice reached what seems to be the limit of its retreat – the polar regions and those places where glaciers are still to be seen today, on the Scandinavian mountains and in the Alps.

So when the ice of the latest Ice Age – in which our own age is only an interlude – began to retreat from England, there were plants here, and animals living on them, and men hunting those animals and living on the plants. They were men of the Palaeolithic, the Old Stone Age, men such as we are, with brains as powerful and hands as skilful as our own, differing from us only in one important respect – they were poorer by ten thousand years of learning than we are.

When the ice had gone from all but the highest hills in the far north the country was at first almost treeless, a watery landscape of bogs and marshes and great shallow lakes and rushing rivers and stone scoured or shattered into scree by the retreating ice. And lichen and grass, which provided food for the herds of wild horse, reindeer, bison and even mammoths.

But meanwhile the forests were advancing northwards across Europe from those enclaves of warmth where they had survived the long winter of the Ice Age. You can still see Tertiary flora in some of those places: for example the *Parrotia* forests on the hills west of the southern Caspian; and the great laurel forests on the island of Gomera in the Canaries.

But how can a forest move? It is very simple really. For tens of thousands of years the millions of seeds released every year in autumn

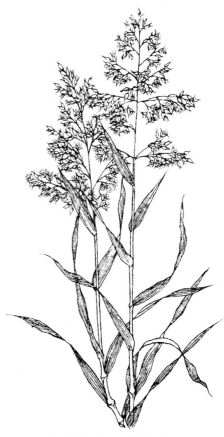

Yorkshire fog, *Holcus lanatus*

by the trees, shrubs and herbs surviving in the warm enclaves south of the worst influence of the ice fields could not germinate beyond an invisible but definite limit at which the conditions became too harsh for them. But as the ice withdrew, as year after year, decade after decade, century after century, the summers became warmer, and the winters less cold, that frontier shifted northwards, and a few of the millions of seeds which fluttered to the ground ahead of the vanguard of the trees each autumn germinated and gave rise to saplings, which became trees bearing seeds in their season. Do not imagine a line advancing like the front line of an army; the wind carried seeds in some places far ahead of the body of the forest, to grow into isolated outposts of the army, to drop their seeds around them in their turn, and so create islands of trees in advance of the main body.

It was not a cheerful landscape; pine forests are great killers of undergrowth and it must have been a dark and monotonous England that became, in due course, one great pine forest. The herds of game moved north away from the trees. The hunting communities moved with them – they had no option – and a new kind of people, men of the Middle Stone Age, the Mesolithic, moved in to occupy the sea-shores, river banks and such open spaces as they could find, and live on small game, berries, nuts, grass seeds, and the roots, shoots and leaves of herbaceous plants.

How did all these plants and animals and men get here, though, since we know Britain is an island separated from the Continental land-mass by enough sea to form an impassable barrier to animals, to boatless men – though they were not long boatless – and to the movement of most plants?

First, at the beginning of this epoch Britain was not an island; it was joined to the Continent in the east by low land separating the North Sea from the Channel, so that Britain was a roughly T-shaped peninsula. In the course of that epoch the sea began to break through. But at first the Channel was narrow and shallow, no serious barrier; its widening and deepening was a work of time, and it took many centuries to reach its present size. The time came when the sea was, indeed, a barrier – one of the reasons why the Continental flora is richer than our own. But seeds cross water, even salt water, unimpaired, in many ways: some seeds are winged and glide great distances; some can float unharmed for long periods of time; some remain unharmed in the digestive tract of birds after the fruit which contained the seed has been digested, and are voided, to germinate where they fall.

Relief from the dark pines was on the way. The deciduous trees – those broad-leaved trees which had solved the problem of surviving the winters by shedding their leaves and going into dormancy – had begun to move in behind the pines before the land bridge joining Britain to Europe had been swept away by the North Sea. A very large number of lesser woody plants, herbaceous plants, annual and perennial, and bulbs were also established here while England was still a part of the mainland.

There was a strange, slow battle between the deciduous and coniferous trees. I have said that the landscape following the retreat of the ice was

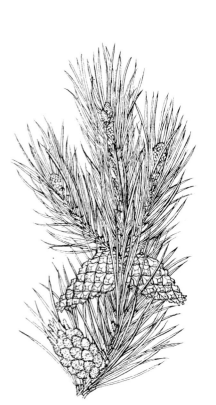

Scots pine, *Pinus sylvestris*

treeless; and so, at first, it was. And if I then went on to the great pine invasion it was because, with the long time-expanses foreshortened by time-perspective, that invasion can be perceived as dramatic, a drawing of a curtain of darkness over England. But with or before the first of the pines came birches and willows – not many, perhaps, and not, on the kind of time-scale we are working on, for long. For although the birches and willows were not exterminated by the pines – they are, after all, still with us – they were restricted and constrained.

The deciduous 'restoration' was perhaps started by those birches and willows so ancient in the English scene, but it was taken over by oak, ash, elm, lime and last of all beech – which remained rare in England until about 1000 B.C. There is a theory that the widening of the sea barrier between England and the mainland favoured the deciduous trees against the pines: certainly something did. The endemic trees of continental Europe north of latitude 51°N are still pines, they still cover much of northern Europe. But – as will appear in the chapters to follow – no pine has been native in England within historical times. Why this difference? As Britain became an island, its climate became milder than that of the Continent; in addition, England, and especially the west coast, had the benefit of the Gulf Stream to warm it. And it was this mildness, favouring the broad-leaved trees, which hastened the retreat of the pines.

So now you have to imagine an England of mixed oak forest wherever the soil was deep and heavy enough, almost everywhere but the gravel terraces in river valleys, the chalk downs of Sussex, Dorset, Norfolk, Lincolnshire and Yorkshire, the limestone hills, the marsh-lands, and the high moorlands.

Until this point – the point at which the pines had gone and the oaks stood in their place – the people of England had been hunters and gatherers of wild vegetable foods. They were thus as much an element of the ecological community as the animals and birds they hunted, the fish they caught and the plants they relied on for nourishing leaves, roots, or fruits. But the way of life far to the south and east, in west Asia, the Nile Valley and the Mediterranean islands, was changing. There, for thousands of years, men had been developing the arts of deliberately growing the vegetable foods they needed; and of herding and breeding instead of hunting the animals which provided them with meat, hides, fibres, bone, hoof and horn – the raw materials of industry.

These arts spread from the centre of invention and development outwards, by three processes. First, a kind of social osmosis, by means of trade, whereby the practice of farming penetrated to the peoples on the periphery of the farming centres, so that those centres grew steadily larger. Secondly, by emigration: emigration was forced on a proportion of the early farmers by pressure on space. Farming, with its relatively high and relatively stable production of food and raw materials, so improved the human condition that the rate of population increase became progressively faster, enormously faster than it had ever been in hunting and gathering economies. This led to the pressure on land, which forced emigration on a proportion of the population, a 'hiving-off' like that of bees swarming from an overcrowded hive. These emi-

Pedunculate oak, *Quercus robur*

15

Mossy saxifrage, *Saxifraga hypnoides*

Great panicled sedge, *Carex paniculata*

grants did not go further than they had to; they came to unoccupied country suitable for their purpose, they settled; but in due course the same situation arose and some had to move on again. And as they took with them on their migrations both their increasingly domesticated farm animals and their seed-corn, they automatically spread both the techniques and materials of farming further and further from the original centre.

Now we come to the third and perhaps most important cause or process of migration: in the original centres of the great agricultural proto-civilizations, Mesopotamia, for example, and the Nile Valley, there was a natural annual renewal of soil by the deposit of flood silt; this made movement unnecessary, and it is why the first appearance of real settlement and urbanization occurred in those centres, rather than elsewhere. But when the farming techniques were carried to other areas as a consequence of pressure on land by rising population, a grave and unforeseen problem arose: exhaustion of pastures and of soil fertility. This problem was, in due course, to be solved by the invention of manuring. But until then people were driven to seek new lands by the impoverishment of the old ones.

It took three thousand years at least, from the time of the earliest farmers in the Middle East, for the Neolithic (New Stone Age), migrant-farming culture to reach England; the farmers must, I suppose, have been hard pressed, because by then Britain was an island and they had to carry their cattle and sheep, as well as themselves and their seed-corn, over the Channel in boats. The first of these Neolithic farmers arrived about 5500 years ago. They had heavy stone axes, boats, digging tools, sewing tools of bone; they had cattle, sheep, pigs, dogs and seed-corn, probably both wheat and barley. But above all they had a purpose: to clear land and make farms – and villages, too, of a sort. They differed from all their predecessors in the land by being no longer one of the manifold and multiform elements of an entirely natural ecology, no longer just mentally clever and manually skilful animals, but outside nature, set, though they might not know it, on taming 'nature'.

Up to this point, then, the flora of England was made by nature. At this point – perhaps about 3500 B.C. – man took a hand and set about remaking it. The impression made by these first Neolithic farmers was not great: their numbers and their means were small. They felled trees and used fire to help them make clearings in the forest. They introduced wheat and barley and probably a few weeds of cultivation into the flora, which henceforth includes cultivated as well as wild plants.

In the course of the next forty-five centuries man transformed the English flora.

Let us look now at what we were given to work with.

First, there were those elements of the flora which had survived the last Ice Age (in the south of England, which was never covered by the ice), and were adaptable enough not to be discouraged out of existence by the increasing mildness of the climate.

Botanists recognize a group of about 270 species of diverse genera and families, whose principal centre is now Scandinavia and the sub-

16

arctic, but which are still to be found in Britain (fewer than 200 of them in England). Many of these plants do survive much further south in Europe, but only as alpine plants, at an altitude where more or less subarctic conditions prevail. Many of these species are plants which only grow in peat soils, others are to be found growing only in wet bogs or in tundra-like conditions. Their English representatives are concentrated in the northern half of the country, although there are sometimes enclaves of plants belonging to this group on high moors or in mountain country further south.

Fossil remains of many plants in this group of species have been identified in late glacial deposits. This, of course, is evidence that they were here before the glaciers crept over England; or that they arrived some time during the period of the four glaciations, perhaps during one of the milder interglacials. Here are a few examples: the insectivorous droseras, which grow in bogs; some gentians; the pretty marsh pea, *Lathyrus palustris*; a large number of the sedge species; the *Vaccinium* – bilberry – which grows on our high moors even in the southwest. Among the absolutely certain pre-glacial survivors are ajugas and alchemillas, more of the sedges, a considerable number of saxifrages; one *Trollius* species and the rare Teesdale violet, *Viola rupestris*.

So, as the ice withdrew, the flora of England was probably composed of a couple of hundred species of herbaceous perennial and annual plants and a few species of small trees or shrubs, a very large proportion of the plants being water-loving kinds such as sedges. All the rest of the species in our flora reached us in the post-glacial period, that is to say during the last eight thousand years or so, and naturalized themselves as the climate became milder. As I am here using the word 'naturalized' in a way which no scientist would approve of, it will be as well to discuss why I do so.

When we make use of this word to refer to the process whereby plants which were alien have become native, we usually have in mind those plants which, introduced by man for cultivation, have 'escaped' from the farm or garden and established themselves in the wild: martagon lilies, for example, or *Mahonia aquifolium*. But for my part I can see no justification for making a distinction between the naturalization of plants which were deliberately introduced into England by man and those which reached us by accident, as it were, including accidental introduction by man. The whole process of the making of the English flora since the end of the glacial period between ten and eight thousand years ago has been one of 'naturalization' of returning former natives or of aliens.

There is a large group, or several groups, of plants now native here which cannot be linked to a particular geographical area because the species in question are so widely distributed that they could have migrated originally from a number of places. There is a large number of species of plants found wild in England which are also found wild all over Europe, Asia, North America and North Africa, so obviously we cannot say where they came from, after the ice withdrew; some of them may even have been here before the Ice Age and survived it. There is another group whose members are not found in North America but

Marsh gentian, *Gentiana pneumonanthe*

Teesdale violet, *Viola rupestris*

Bilberry, *Vaccinium myrtillus*

Martagon lily, *Lilium martagon*

Welsh poppy, *Meconopsis cambrica*

are common to the whole of the northern temperate zone of the Old World – from Britain in the west to Japan in the east; so, again, attribution of a geographical source is impossible. There is a third group whose origin can be narrowed down to Europe, since its members become increasingly rare as we move north-east and are not found in Asia nor, with a few minor exceptions, in North Africa. There is also a small group called 'endemic' – species found nowhere but in Britain and Ireland, though we can find their close relatives elsewhere. But there are also groups which can be traced to a relatively restricted geographical source.

There are thirty-eight species of plants wild in England which can be traced to a source in the Mediterranean region. These must have worked their way northwards across France or up the Atlantic coasts of Portugal and Spain, as, year after year, the climate became milder and their potential habitat extended. These species established themselves in the Channel Islands and all along the south coast from the Isle of Wight to Cornwall; and they become scarcer and scarcer the further north you go. Some of these migrants did not quite reach England: *Arbutus unedo*, the strawberry tree, for example, did not get beyond southern Ireland. Among the most attractive species in this group are one of the forget-me-nots, one of the centaureas, the purple gladiolus and a ranunculus. The list includes silenes, lavateras, and salvias. Most of the species flourish only in rather dry, sunny places and are, therefore, marginal in Britain.

Next we have a list of eighty species which can be connected with south-west Europe and must have made their way northwards from that region between six and seven thousand years ago. The list includes pimpernels, spurges, rock-roses, St John's wort, a ranunculus – not the same species as the 'Mediterranean' one – the lovely little *Scilla autumnalis*, which I have found only in Devonshire, a number of clovers and the familiar and beloved holly. In the case of each of these groups, the plants named – and the same is true for the rest of these lists – 'settled' here in the parts of England which, in soil and climate, most closely resemble the source regions.

Next there is a group of some eighty-five species which can be connected geographically with the Atlantic coast of continental Europe, from Holland in the north to Portugal in the south. At a guess, they were perhaps among the first post-glacial arrivals. Among them are the wild cabbage, two of our heaths, *Erica ciliaris* and *Erica tetralix*, the autumnal crocus, the bluebells, some of the orchids, the yellow 'Welsh' poppy (*Meconopsis cambrica*), and gorse.

A group of nearly 130 species can be connected with an area extending from Central Europe to North Africa but is not specifically Mediterranean. Most of the English representatives of this group are found in south-east England. North-west of a line from the Bristol Channel to the Wash there must have been and in a measure still is a climate-*cum*-soil barrier which a majority of these species could not pass; about twenty of them succeeded in doing so. This long list of species includes box, traveller's joy, *Daphne laureola*, snowdrop, some orchids for the most part rare in England, the woad (*Isatis tinctoria*) of the ancient Britons, honeysuckle, mint, some of the poppies, a viburnum and the lesser periwinkle. Concentrated chiefly in the south-east of England, but with a small number of its members extending further into the north-east, another group of plants can be connected quite clearly with central and east Europe and western Asia and south Russia, virtually, then, with the great Eurasian steppe. The question is, were they Ice Age survivors or re-colonizers of their English territory? Among these are the pasque flower, *Pulsatilla vulgaris*; the hornbeam – certainly a re-colonizer; a number of orchids; the snake's head fritillary, *Fritillaria meleagris*; the beech and both our own species of oak; the dusky cranesbill; two rose species; and one of our elms.

*   *   *

Until this point we have been talking chiefly about angiosperms, that is, flowering plants, which produce seed enclosed in an ovary, and also the very few gymnosperms, plants which produce seed not enclosed in an ovary, such as, for example, the conifers. But there is a large class of plants belonging to the Filicineae, which produce no seeds but propagate themselves by means of spores. In other words – ferns. And they have many representatives in our flora.

There are about forty species of ferns native to Britain, perhaps about thirty-five native to England, but a much larger number of varieties and inter-generic hybrids.

It is probable that a large majority of these ferns are pre-glacial survivors; but by no means all of them. It is, for example, difficult to conceive that the maidenhair fern, which is confined in England to the mild south-western peninsulas, survived the cold of the last glaciation. Nor are the filmy ferns, *Hymenophyllum*, hardy. On the other hand some English ferns are ruggedly hardy, even so majestic a plant as the royal fern, *Osmunda regalis*.

At all events, we have to add to the picture of our flora as it was at the beginning of history a considerable number and wide variety of ferns.

*   *   *

What, after all this, does the picture look like now?

According to the most respected authorities, the number of species which are what botanists call native – that is, which got here by agencies other than man, i.e. 'naturally', and for the most part before man was in a condition to do anything about the flora – is about 1500, belonging to about 500 genera of plants and about 100 families.

How does this compare with our neighbours' flora? Very poorly; a single example will suffice: France has four times the number of indigenous species. Why should this be so? Because, as I have suggested elsewhere, at the time when the plant re-population of northern Europe following the Ice Age was in full spate, the opening and progressive widening of the Channel more or less barred any more immigrants. There is, of course, another reason: a considerable number of species flourishing at low altitudes round the Mediterranean's north shore are too tender to get a footing in England – even species introduced by

Woad, *Isatis tinctoria*

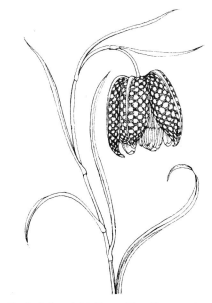

Snake's head fritillary, *Fritillaria meleagris*

Dusky cranesbill, *Geranium phaeum*

Ferns: (1) Hart's tongue fern, *Phyllitis scolopendrium*

(2) Tunbridge filmy fern, *Hymenophyllum tunbridgense*

(3) Rusty-back fern, *Ceterach officinarum*

(4) Maidenhair spleenwort, *Asplenium trichomanes*

(5) Beech fern, *Dryopteris phegopteris*

(6) Wall rue, *Asplenium ruta-muraria*

(7) Hard fern, *Blechnum spicant*

(8) Maidenhair fern, *Adiantum capillus-veneris*

(9) Prickly shield fern, *Polystichum aculeatum*

(10) Common polypody, *Polypodium vulgare*

(11) Lady fern, *Athyrium filix-femina*

(12) Royal fern, *Osmunda regalis*

man from the South of France to England and doing well here in gardens, e.g. rosemary and lavender, have never established themselves here in the wild.

On the other hand we have a number of species, the 'endemic' species of England, which are found nowhere else in the world. How has this come about? Well, it is something which tends to happen on islands; the Canaries, for example, have a remarkable endemic flora. Those 'pressures' which motivate evolutionary change are different from and in some respects more limited than those of continental territories. Moreover, a mutant has a better chance of being relatively 'segregated' and so of establishing itself unchanged.

*　　*　　*

But 1500 is not the sum of plant species growing in the wild in England. There are, in our flora, at least 480 and possibly over 500 plants which were introduced deliberately or accidentally in historical times and have established themselves as members of the flora. The total of species wild in Britain is, then, about 2000, to which we must add a further 2000 or so commonly cultivated species.

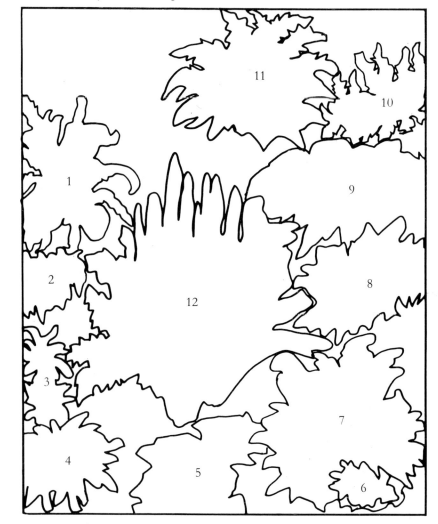

Scientific Names of Plants mentioned in Introduction and Chapter One

COMMON NAME	LATIN NAME	FAMILY
Autumnal crocus	*Crocus nudiflorus*	Iridaceae
Beech	*Fagus sylvatica*	Fagaceae
Bilberry	*Vaccinium myrtillus*	Ericaceae
Bluebell	*Hyacinthoides non-scriptus. H. hispanicus*	Liliaceae
Box	*Buxus sempervirens*	Buxaceae
Bugle	*Ajuga* spp	Labiatae
Cabbage, wild	*Brassica oleracea*	Cruciferae
Clover	*Trifolium* spp (some)	Leguminosae
Cranesbill, dusky	*Geranium phaeum*	Geraniaceae
Daphne	*Daphne laureola*	Thymelaeaceae
Ferns (many)	(several genera)	
Fritillary	*Fritillaria meleagris*	Liliaceae
Gentian	*Gentiana* spp (some)	Gentianaceae
Globe flower	*Trollius europaeus*	Ranunculaceae
Gladiolus, purple	*Gladiolus illyricus*	Iridaceae
Gorse	*Ulex europaeus*	Leguminosae
Health	*Erica ciliaris, E. tetralix*	Ericaceae
Holly	*Ilex aquifolium*	Aquifoliaceae
Honeysuckle	*Lonicera periclymenum*	Caprifoliaceae
Hornbeam	*Carpinus betulus*	Corylaceae
Lady's mantle	*Alchemilla vulgaris, A. alpina*	Rosaceae
Mallow, tree	*Lavatera borea, L. cretica*	Malraceae
Oak	*Quercus robur, Q. petraea*	Fagaceae
Orchids	(several genera)	Orchidaceae
Pasque flower	*Pulsatilla vulgaris*	Ranunculaceae
Pea, marsh	*Lathyrus palustris*	Leguminosae
Periwinkle, lesser	*Vinca minor*	Apocynaceae
Pimpernel	*Anagallis* spp (some)	Primulaceae
Poppy, Welsh	*Meconopsis cambrica*	Papaveraceae
Poppy	*Papaver* spp (some)	Papaveraceae
Rock rose	*Helianthemum* spp (some)	Cistaceae
St John's wort	*Hypericum* spp (some)	Hypericaceae
Salvia	*Salvia* spp	Labiatae
Saxifrage	*Saxifrage* spp (many)	Saxifragaceae
Sedges	*Carex* spp	Cyperaceae
Silene	*Silene* spp (some)	Caryophyllaceae
Snowdrop	*Galanthus nivalis*	Amaryllidaceae
Spurge	*Euphorbia* spp (some)	Euphorbiaceae
Spurge laurel	*Daphne laureola*	Thymelaeaceae
Squill, autumn	*Scilla autumnalis*	Liliaceae
Sundew	*Drosera rotundifolia, D. intermedia, D. anglica*	Droseraceae

Box, *Buxus sempervirens*

Meadow clary, *Salvia pratensis*

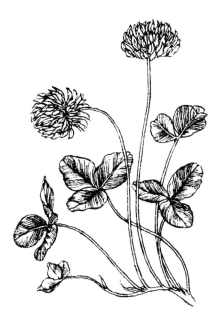

White clover, *Trifolium repens*

Crocus, *Crocus nudiflorus*

Spurge laurel,
Daphne laureola

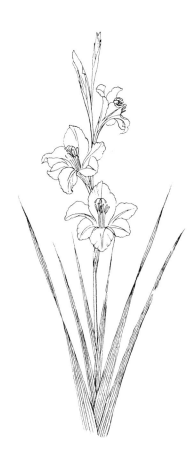

Purple gladiolus,
Gladiolus illyricus

Water mint, *Mentha aquatica*

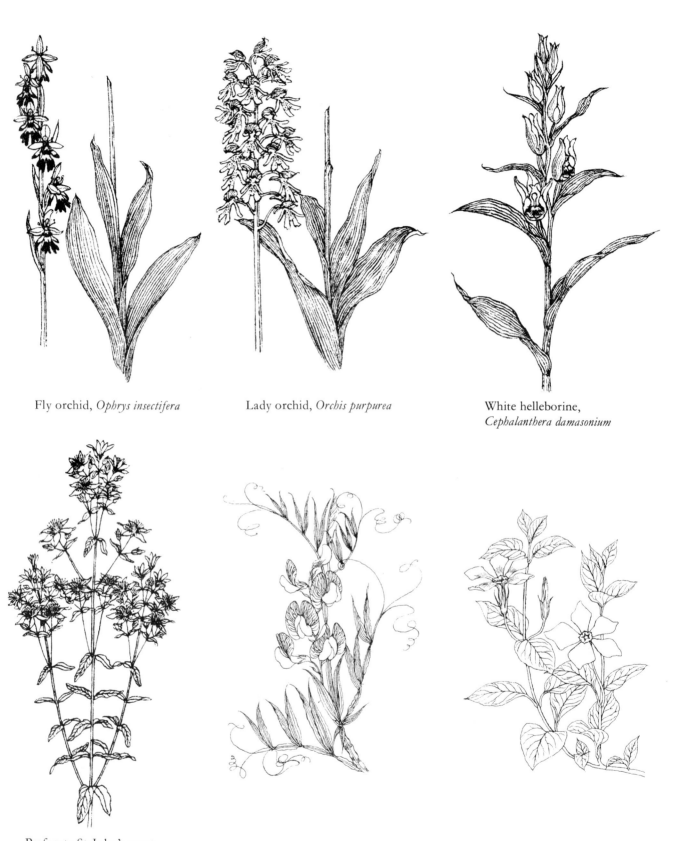

Fly orchid, *Ophrys insectifera*

Lady orchid, *Orchis purpurea*

White helleborine,
Cephalanthera damasonium

Perforate St John's wort,
Hypericum perforatum

Marsh pea, *Lathyrus palustris*

Lesser periwinkle, *Vinca minor*

24

Scarlet pimpernel, *Anagallis arvensis*   Sea wormwood, *Artemisia maritima*   Wood spurge, *Euphorbia amygdaloides*

COMMON NAME	LATIN NAME	FAMILY
Traveller's joy	*Clematis vitalba*	Ranunculaceae
Violet, Teesdale	*Viola rupestris*	Violaceae
Wayfaring tree	*Viburnum lantana*	Caprifoliaceae
Woad	*Isatis tinctoria*	Cruciferae

Alder, *Alnus glutinosa*

Crack willow, *Salix fragilis*

Chapter Two
Trees

IF you walk or drive through the English countryside you get the impression that all the trees look perfectly 'natural', that they 'belong'. And, in a sense this is true, just as it is true that people of Norman or Danish or Dutch or Saxon origin are as 'English' as the earliest inhabitants of Britain. But because we are trying to discover how trees which were not here when the prehistoric inhabitants of Britain were looking at the land they lived in have changed the look of England's flora, we make a distinction between our primeval trees and those we have now.

We will suppose, then, that we are taking a long day's drive to look at trees.

Over there, on the left, is some rather marshy land beside a river; there are two kinds of trees, none very big, some with dark, rough-looking, oval leaves, the others with much lighter, silvery, long, narrow leaves: **alders** and **willows**. There are many species of both in the world; one of the alders has been native here since before the Ice Age, which it survived, so we can reckon it among the oldest inhabitants. As for the willows, they were among the first trees to reappear in England as the ice withdrew further and further north; and not, in their case, just a single species – four are among England's really ancient tree species – the white willow, the round-eared willow, the sallow, and the species which were the ancestors of the cricket-bat willow.

On our right as we pass through a village there is a front garden consisting of a lawn surrounded by an evergreen hedge, neatly clipped, with a magnificent cedar in the middle and a group of birches, their trunks almost white, in one corner. The hedge is **boxwood** which, in warm weather, has a pleasant musky smell. It is another ancient native and there were once, when man was still a hunter in Britain, box forests, slender, graceful trees growing on the limestone hillsides as there still are, for instance, in France and Iran. The **birch** tree, too, is ancient here, another of the first trees to colonize the land when the Ice Age ended; in fact, it is possible that the birch may have survived the Ice Age and remained to brave the bitter cold from the warmer days before the ice came.

What about the **cedar**, though? The one in that garden looks like cedar of Lebanon, the tree which provided the timber for building Solomon's temple in Jerusalem about 3000 years ago. The first ever planted in England arrived at about the same time as William of Orange and his Queen Mary Stuart, late in the seventeenth century. There are two other species of cedar. You will probably see specimens of both in a day's drive; but your great-great-grandfather could not have done because one, the deodar, did not reach here from India until 1830 and the other, the Atlantic cedar from north-west Africa, until 1840. So none of the cedars is 'natural' here.

Clear of the village we are driving between tall hedges of **hawthorn**, equally pretty in spring when they are in flower and in autumn when they bear heavy swags of scarlet berries. There are about a thousand different species of hawthorn trees growing in all the northern hemisphere. One or two are among the most ancient of England's trees, including the one used for making hedges.

Away to our left front the land rises to form gently rounded hills covered with fine grass with groups of enormous trees with very smooth bark, almost the colour of an elephant's skin, a great spreading head of leaves, the older leaves almost bronze-green, the young ones very pale green: **beeches**, of course (see p. 11). 'Natural' or not? Well, the beech has not been in England nearly as long as alder, willow and hawthorn, but probably for between three and four thousand years, so it is not a recent immigrant like the cedars. On the whole, let us call it natural.

On the other side of the road the land is flatter. A great expanse of rough grass with some groups of old trees and cattle grazing under them. Among the trees is a group with very tall trunks, leaning eastward, and dark, flat, spreading heads: **Scots pine** by the look of them (see p. 14), and surely perfectly 'natural' in an English scene? As a matter of fact, no, not really; and the name 'Scots' pine gives us a clue. The surprising fact is that no pine, no conifer of any kind has been natural to England since about 6000 years ago. What happened was this: the Scots pine is a very hardy tree which can stand a great deal of cold. When the glaciers which once covered so much of England began to retreat and the climate became warm enough for trees, these pines, which had survived further south, began to move in and there was a time when all England was covered with them. But as the climate got still warmer, so that other, not so hardy, kinds of trees, oak and ash and later beech, could survive in England, the pines, which like cold, began to die out in the south. Meanwhile Scotland, which had been too cold even for pines, was also warming up and the pines could grow there. So by the time man was beginning to increase and multiply and learn how to farm in England, there were no pines left there, but plenty in Scotland. And when, say about five hundred years ago, men began deliberately introducing foreign trees to England one of the first was the pine – from Scotland.

We are now approaching a town which has become one by filling in the spaces between a number of old villages. There is even the village green still there in the first part we come to, with a pond for ducks and geese and one enormous old **ash** tree, so huge that it might be the oldest living creature in England. In fact, though, it is not: it might be a hundred, maybe even two hundred years old. But the ash has been in England since the Ice Age ended and that is about eight thousand years ago. So we are right to feel it is 'natural' in the English scene. There is a fine **holly** hedge round one of the bigger houses, and there were some taller holly trees in the hedge we passed on the way into the town; the

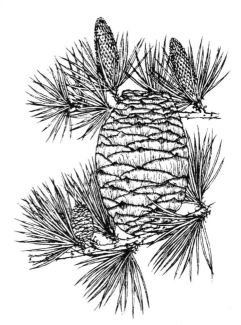

Cedar of Lebanon, *Cedrus libani*

Ash, *Fraxinus excelsior*

Juniper, *Juniperus communis*

hollies have been here as long as the ash trees: there is no doubt about their being natural here. But what about the big **monkey puzzle** tree in the front garden of that important-looking house, which must belong to someone rather rich by the look of it? I suppose that nobody would ever take monkey puzzle trees for natural in England; I mean, they do not *look* natural. And, of course, they are not; despite the fact that you can see them in suburban gardens all over England, they come from Brazil; and the way they got here is odd. In 1793 the famous botanist Archibald Menzies, a member of the explorer ship *Vancouver*'s company, was dining together with the ship's officers with the Viceroy of Peru. Nuts from the cones of the monkey puzzle trees were served as dessert. Menzies put a few in his pocket, planted them in pots back on board ship, and on returning to England had five young monkey puzzle trees.

Our drive is taking us past, if not through, limestone country: those rounded hills with fine 'stands' of beeches were limestone hills, which is why on the slopes below the beeches there were a few small trees rather like miniature firs – **junipers**, in fact. This species is among the primeval trees of England and certainly quite 'natural' here. Gin, by the way, gets its name from this small tree: the Dutch word juniper is *ginever*; juniper berries were used to flavour the spirit, which was therefore called *ginever* – gin for short.

The trees planted along the streets of the towns we are passing through are **planes** (see p. 38), the commonest street trees in the Western world, not only in Britain but all over the northern hemisphere. They grow magnificently in the polluted air of cities – look, for example, at the superb plane trees in Hyde Park, London. The history of this tree is curious. There was a species of plane native to our territory before the Ice Age: it was wiped out by the great cold and it never returned. So, from the time of our prehistoric ancestors who lived here after the Ice Age up to about A.D. 1500, no stay-at-home Englishman had ever seen a plane tree. Then a species from south-east Europe was introduced, and later a species from America. By natural accident these species cross-bred and seedlings were found which were hybrids between the two. Those were the first 'London planes', ancestors of all the town and city plane trees in Britain and North Europe.

If our tree-watching drive were made in the month of May we should notice the tops of tall, slender trees standing out above the tree-tops of certain coppices, and covered very prettily with white flowers; these are England's wild cherry trees: the old name is **gean** (a corruption of the French *gigne*). It is, like those other members of its genus, the **bird cherry** and the **blackthorn**, among our primeval trees, a true native. In flower at the same time as wild cherry are the huge old **horse chestnuts** on a green outside the gates of the park of a great country house which is the next feature of the landscape we are looking at. Natural though they look, as if they had been growing here for ever, they are foreigners and, comparatively speaking, newcomers. In 1576 the great Flemish botanist Charles de l'Écluse, better known as Clusius, was working in Vienna for the Emperor of Austria. A friend of his, the Austrian ambassador to the Sultan of Turkey in Constantinople, sent him seeds of a lovely flowering tree which grew in Turkey; Clusius planted these

Bird cherry, *Prunus padus*

'conkers' and raised the trees, which came to be known as horse chestnut because the nuts are so like sweet chestnuts. Seeds were sent to botanists in all the European capitals and the first horse chestnut was seen in England in 1600.

How about that coppice, over on the right near that farmhouse, of **sweet chestnut**, grown with a dozen small trunks instead of one big one, to provide stakes for young fruit trees, wood for fences and hurdles? The sweet chestnut is no more truly natural here than the horse chestnut, but it has been here much longer and there are so many magnificent old chestnut trees in old country house parks, old avenues and even growing wild in woods that it is hard to believe the tree is not a native. It was first planted in England by the Romans, for its nuts, its shade and its timber, which is almost as good as oak.

Talking of **oak**, we have passed hundreds of them in the course of our drive – you can hardly help doing so in England. There are splendid old specimens in parkland, smaller ones in the hedgerows. There are hundreds of species of oak in the world, both deciduous (dropping their leaves in late autumn) and evergreen; many have been introduced here in the last two centuries, including the beautiful evergreen or **holm oak**, which you usually find at its best near to the sea. But some species are true naturals here; in fact, after the pines had retreated north into Scotland, the most common trees, covering almost all England except the chalk hills and downs, were oaks. It was in oak forest that our ancestors hunted the deer and later kept great herds of pigs, which lived on the acorns; it was oak forest the first farmers in England had to clear to make room for crops; it was oak which was used to build houses and ships. And the reason that the oak forests disappeared was that the wood was used to stoke furnaces for smelting iron until the iron-smelters learnt how to use coal.

Here and there we find bits of oak forest still surviving. We shall leave the car and take a walk in one. There are other kinds of trees growing among the oaks and presently we come to a rise, with four enormously stout, low, spreading trees which look like conifers and have nearly black foliage (see p. 38): **yews**, more familiar in churchyards than in woods. The yew is primeval here. Why is it so often found growing in churchyards? Some say because the Romans used it as a temple tree, as a substitute for their own cypresses; others that it is because the trees look mournful; yet others that it was to have handy a supply of yew timber for making bows (although most of the yew for bows was imported from Spain).

Resuming our drive we find the road running between solid blocks of tall, slender **spruce** trees, a whole forest of them, but obviously a forest planted not by nature but by man. We shall probably see more of such forests in the course of the drive. Why so many of these spruces, and not only the common one but several other spruce species? Because this species is now the most important source of timber we have – timber for building and for making wood pulp, which is the raw material the paper for newspapers is made of. But this tree is a foreigner, first introduced here from northern Europe in about A.D. 1600.

We have already noticed Scots pines, but as we drive it is impossible

Horse chestnut, *Aesculus hippocastanum*

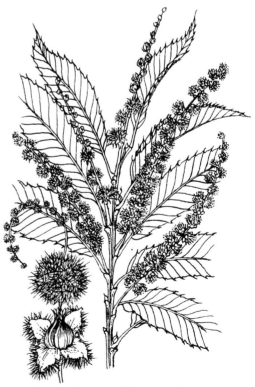

Sweet chestnut, *Castanea sativa*

Two imaginary landscapes showing, *above*, trees long established in the English countryside (among them are Scots pine, beech, holly, juniper, birch, hawthorn, poplar, ash, elm, alder, sallow, willow, oak, yew, box and wild cherry)

Below, a similar scene filled in with 'imported' species (including chestnuts, cypresses, maples, redwoods, a monkey puzzle, a tulip tree and a magnolia)

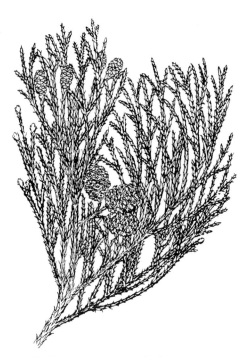

Wellingtonia, *Sequoiadendron giganteum*

Swamp cypress, *Taxodium distichum*

not to have noticed, here and there, usually in big gardens or parks, a lot of trees which are clearly **pines** but very different from Scots pines; and a lot of **fir** trees, with much shorter 'needles' than the pines. Not a single one of these is 'natural' here. During the eighteenth century species of pines were introduced for their beauty, or their timber, or both, from west Asia and from both north and south Europe; probably the earliest on the scene was the **stone pine** from the Mediterranean region. Then, in the nineteenth century, came a whole lot more species from North America, China, Japan, Manchuria, Siberia and the Himalaya mountains. The planting of pine trees quite changed the look of England, which before them had had no big evergreen trees in its landscape.

Those **poplars** beside the road, planted to screen an orchard from the wind, are introduced, not native. But we do have poplar species which are primeval, though now less commonly seen than the foreign white poplar, black poplar, Lombardy poplar and aspen poplar.

We shall stop for a few minutes now to take a look at what used to be the private park of a ducal mansion but is now a public park on the outskirts of a city. The last three dukes, grandfather, father and son, were great lovers of trees and planted thousands. The most striking of these, right in front of us, tall, conical, dark and very stately, are **Wellingtonias** and **redwoods**. None of these had been seen in England until about 150 years ago, when seeds from their cones were brought from California; those specimens are a little over 100 years old, but they are only babies, for many trees of their kind are more than 3000 years old in their native California, attaining a height of more than 350 feet and a girth of nearly 100 feet. You might think them to be the oldest living things in the world but they are not. There are some pines (species *Pinus aristata*) still alive in South California which are 4900 years old and were big trees before the pyramids in Egypt were built.

There is a river running through the park and some rather water-logged meadows beside it, and there, among the familiar willows and alders, some tall and graceful conifers with foliage of a lovely light, fresh green. They are **swamp cypresses** and look perfectly natural in that scene; but in our sense, they are not. There were no such trees in England until the first were introduced from America in the seventeenth century. Between them and where we are standing is a group of smaller trees which you can be forgiven for thinking are also swamp cypresses, only younger, so alike are they at first glance. But they are **dawn redwoods**, *Metasequoia glyptostroboides*, and they have an extraordinary history. Before 1945 these trees were known only as fossils and botanists believed that they had been extinct for many millions of years. But in that year a Chinese botanist found a group of these trees growing in a province of China, Szechuan. In 1946 two Chinese botanists, Doctors Hu Hsen-hsu and Cheng Wan-chun, went to the place where these last survivors of an immensely ancient species were growing, and collected seeds which they sent to the botanical gardens of the Arnold Arboretum in Boston, U.S.A., and to Kew Gardens in London. Seedlings were raised and distributed and now we have these beautiful trees growing in parks and gardens all over Britain and the U.S.A.

Across the park, beyond the Wellingtonias and redwoods, there is a tall hedge of some conifer, beautifully trimmed and clipped so that it looks like a green wall. Again, no native this but an American, **arbor-vitae**, which first reached us in the seventeenth century. Just in front of it and looking rather like it except that it has not been clipped, is a group of **hemlocks**: they first reached us from Canada a couple of centuries ago.

Balancing the hemlocks, on the other side of the landscape within our view, is a small group of trees which, at first glance we would take for beeches. Well, they are a sort of **beech**, *Nothofagus antarctica*, but they come from the southern hemisphere – from Patagonia in fact; they are closely related to our own beeches and were introduced here in the last century. They seem quite at home in England and I expect that in a hundred years or so, they'll be taken for natives.

We drive on: we come to some very flat country with enormous fields of wheat, a few fields of maize, big, handsome farmhouses, fields of peas; beside the road are tall, very handsome trees whose profile is very familiar to all English people. On some of them the leaves of some branches are yellow and withered; it looks as if they are sick, and they are; for these are **elms** and their sickness is Dutch elm disease: shall we lose them all, then? It is not likely but it is possible. We have several species, all primeval, all native to the natural English tree-scape. Two of the species are so English that they are not to be found growing wild anywhere else in the world.

You will notice, if you are looking at trees when you are in a car or walking, that **maple** trees are much commoner near towns, or in gardens, than anywhere else. If you kept a record of every one you see you might get as many as forty different species now well-established in our English scene; but all bar one are foreigners introduced deliberately during the last two centuries from North America and the Far East. Our native field maple is not very common as a big tree, but it is easy to spot in a hedgerow because its small vine-leaf-shaped leaves turn brilliant golden yellow in autumn. One of the maples which has been here for so long that it is often taken for a true native is the **sycamore**. But there is an easy way to be sure that it is not. Its winged seeds germinate and grow so readily that the seedlings are troublesome weeds in any park or garden where sycamore has been planted. If this species had been in England before there were farmers and gardeners to pull up the seedlings, we should certainly have whole forests of sycamore; but we have not, so the species cannot have been in England much before the fifteenth century.

We have been climbing steadily for some time and we are now driving between stretches of moorland, with heather and bracken and very few trees; the trees are **birches** and we know they are English natives. But are they? We have only one native species; but in 1736 someone introduced the black birch from North America; in 1746 came the paper-bark birch and the poplar-leaved birch, also from North America. More and more birch species were introduced from all over the northern hemisphere in the nineteenth century. So, when we see birches it is by no means certain that they are the ancient English ones until we have taken

Western hemlock, *Tsuga heterophylla*

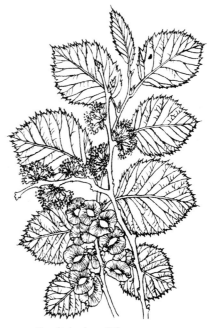

English elm, *Ulmus procera*

Field maple, *Acer campestre*

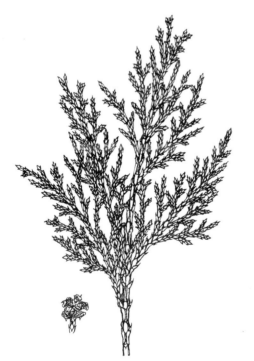

Leyland cypress, *Cupressocyparis leylandii*

a close look at them.

Since our drive is an imaginary one there is no limit to the mileage we can do in a day. I shall take you to look at a garden in Devon which used to belong to me. It is a big one, about three acres, and full of big trees; yet the only ones really native are the oaks, for the pines are foreigners and there are two kinds of tree we have not yet seen anywhere in the course of this drive. They look natural enough, but they are not; in fact they have no ordinary English names. One species is represented by a group of four trees with slender black trunks, about thirty feet tall, the leaves silvery-green with curiously wavy edges and very shiny. They bear thousands of minute dark maroon flowers followed by seeds so sticky that it is hard to get them off your fingers: they are **pittosporum** trees, which reached here from New Zealand some time in the last century.

The other species is a conifer; its foliage is light green but, though it does not fall in winter, it turns a copper colour in autumn. The trunk is bent over in a curve towards the ground; but it only does this in England, for in Japan, where it comes from, it grows into an immense tree, two hundred feet tall, and lives for well over a thousand years: the name is **cryptomeria** and it looks a bit like a cypress tree. Talking of cypresses, there are two in this same garden so enormous in height and bulk and so ancient looking that you would think they had been here since the first Queen Elizabeth was a girl. They are **Monterey cypresses**, and, far from being natives, were not introduced here from California until 1840, though now they are as common as hawthorns, or almost.

There is a cypress – but not to be seen in Devon, they prefer the south-east – which has been here much longer: we shall have to drive into Kent to see them at their best. It is the **Italian cypress**, and was introduced about 1600. It is called 'Italian' because it is common in Italy and we had it from there; but as a matter of fact it is never found wild in Italy. I have seen the place it originally came from, in the mountains west of the Caspian Sea in Iran.

There is a cypress which is unquestionably English but the species is not yet fifty years old. Here is the story: in a private park where the Monterey cypress was growing as a group of trees quite near to another cypress species introduced from America, the Nootka cypress, some seedling cypresses sprang up which turned out to be hybrids between those two species. It is now called the **Leyland cypress** and is the fastest growing and one of the most beautiful evergreens we have.

On the way back from that Devon garden we pass a number of places – gardens or parks – in which we notice some trees so strangely unlike those we are familiar with that, although they are growing well, we do not for a moment take them for natives. Their bark is marbled white, green and buff, and in places it is peeling off in great strips; they have two quite different kinds of leaves; on the young shoots the leaves are oval, on the old shoots long and narrow and rather dismal-looking; they are silvery-blue-green in colour and have a strong medicinal scent: they are **gum** trees from Australia or New Zealand and have been here only for about fifty years.

As we get further east in the course of our journey back, we drive along one stretch of road where the hedgerow trees are neither oak, nor ash nor elm, but **walnut** trees. It is a relief to see any nowadays; so many walnut trees have been felled because of the high price of their timber that they are now quite rare. The walnut is not a native, natural though it looks in an English setting. There is no record of the date of its introduction but it was probably the Romans who first planted walnut trees here. As a matter of fact, though, it is not native to Italy either; the Italians first had it from the Greeks who had it from Iran, where it is native. My guess is that the first walnut trees were seen in Britain about A.D. 300, at the same time as the first grape-vines but later than the first fig trees.

Now we pass a plantation of very graceful coniferous trees with straight trunks and rather light green foliage, next to a house with a lawn beside it in which stands a single tree which we have not noticed before. Those conifers are obviously not pines, firs or cypresses: they are **larches** and most people would call them native English trees. But, once again, they are not. The first larches in Britain were brought from the Alps and planted at Dunkeld in 1738. So good is their timber that very soon larches were being planted in a big way in many parts of England, until they were more numerous and more important as timber trees than the native oaks, in fact, the most important of all the timber trees, a position they held until displaced by the spruces. This European larch is naturalized here and often found growing wild, yet 250 years ago it was an exotic novelty. As for that specimen tree on the lawn, it has broad, light green leaves shaped like those of a maidenhair fern. It is a **maidenhair tree** and it has a curious history. It was introduced into England from China about 1780 or a little earlier: the young trees for planting had come from a Chinese tree-nursery. The only place these trees were ever seen in China was in temple gardens, and nobody had ever seen a wild maidenhair tree. Nor has anyone ever done so to this day; the species is extinct in the wild and only survived because it was a temple tree. But it has a very long history. It first appeared about 150 million years ago. Some fossil leaves from rocks of that time are identical with those of the living tree.

The next big house with its lawns and specimen trees which we come to has a number of ornamental trees worth looking at. On one wall of the house, as tall as the eaves and almost obscuring some of the windows, is a tree with black bark and big, laurel-like leaves, dark green and shiny on one side, light brown and felty on the underside. It has one or two huge, goblet-shaped, creamy flowers. It is a **magnolia**, introduced from America in 1720. No magnolia had been seen here before then, but since its introduction over thirty different species have been introduced. This house has two lawn trees, one very tall and spire-shaped – a **tulip tree**, another American immigrant to England – it first arrived in the sixteenth century. The other, lower but broader, a great spreading head of large leaves with a rough surface: **mulberry**.

There are two kinds of mulberries, the white and the red. White mulberries, brought by the French – who had them from the Italians, who had them from the Byzantine Greeks, who had them from the

Walnut, *Juglans regia*

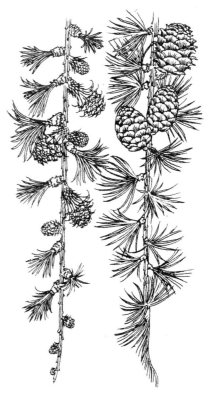

European larch, *Larix europaea*

35

Maidenhair tree, *Ginkgo biloba*

Persians, who had them, some thousands of years ago, from China – on the orders of James the First because he wanted Britain to have a silk industry and the leaves of the white mulberry are what silk-worms feed on. The black mulberry, the kind which has red fruits which are delicious when perfectly ripe, was first planted here in 1548.

*   *   *

What conclusion about the change in the forest-tree section of the English flora must we come to as a result of our imaginary drive? That there are, in our tree scene, about fifteen primeval genera. To them some two hundred generations of men inhabiting this country have added about thirty more genera. But that gives little idea of how much we have, by our own efforts, enriched the diversity of trees in England. For in most of the new genera introduced we have brought in many more than a single species – in some cases five, ten, twenty, even two score.

If the transformation wrought by this imported wealth of alien trees has been chiefly in the, as it were, ornamental landscape, a few species, chiefly evergreen conifers, have changed even the woodland aspect of large parts of England.

Try to imagine a landscape, an England, devoid of pines, firs, cypresses, spruces and cedars, and you will see what I mean.

Tulip tree, *Liriodendron tulipifera*   Holly, *Ilex aquifolium*

H   hybrid   N   native
I   introduced   Nt naturalized
        spp   species

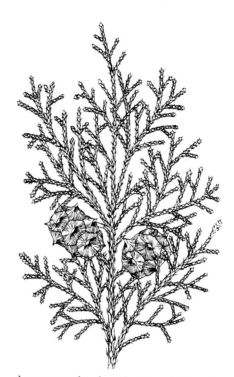

Japanese red cedar, *Cryptomeria japonica*

ORIGIN		GENUS AND SPECIES	FAMILY
N	Alder	*Alnus glutinosa*	Betulaceae
I	Arborvitae	*Thuya occidentalis*	Pinaceae
N	Ash	*Fraxinus excelsior*	Oleaceae
N	Beech	*Fagus sylvatica*	Fagaceae
I	Beech, Southern	*Nothofagus antarctica*	Fagaceae
N	Birches	*Betula nigra*	
		Betula papyrifera	Betulaceae
		Betula pendula	
		Betula populifolia	
N	Bird cherry	*Prunus padus*	Rosaceae
N	Blackthorn	*Prunus spinosa*	Rosaceae
N	Box	*Buxus sempervirens*	Buxaceae
I	Cedars	*Cedrus libani*	
		Cedrus deodara	Pinaceae
		Cedrus atlantica	
Nt	Chestnut, horse	*Aesculus hippocastanum*	Hippocasta-naceae
Nt	Chestnut, sweet	*Castanea sativa*	Fagaceae
I	Cryptomeria	*Cryptomeria japonica*	Cupressaceae
I	Dawn redwood	*Metasequoia glyptostroboides*	Pinaceae
N	Elm	*Ulmus diversifolia*	
		Ulmus glabra	
		Ulmus nitens	Ulmaceae
		Ulmus procera	
		Ulmus stricta	
I	Fir	*Abies*, many species	Pinaceae
N	Gean	*Prunus avium*	Rosaceae
I	Gums	*Eucalyptus*, several species	Myrtaceae
N	Hawthorn	*Crataegus monogyna*	Rosaceae
I	Hemlock	*Tsuga canadensis*	Pinaceae
N	Holly	*Ilex aquifolium*	Aquifoliaceae
I	Holm oak	*Quercus ilex*	Fagaceae
Nt	Horse chestnut	*Aesculus hippocastanum*	Hippocasta-naceae
I	Italian cypress	*Cupressus sempervirens*	Pinaceae
N, I	Juniper	*Juniperus communis* and others	Pinaceae
I	Larch	*Larix europaea*	Pinaceae

Mulberry, *Morus nigra*

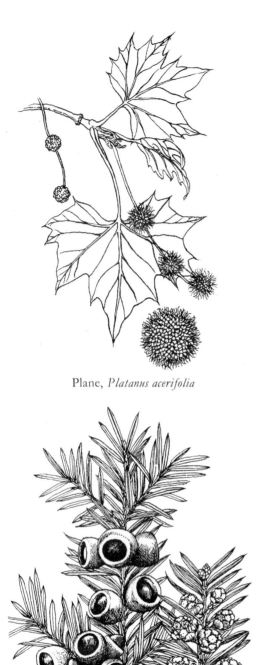

Plane, *Platanus acerifolia*

Yew, *Taxus baccata*

ORIGIN		GENUS AND SPECIES	FAMILY
H	Leyland cypress	*Cupressocyparis leylandii*	Pinaceae
I	Magnolia	*Magnolia grandiflora*	Magnoliaceae
I	Maidenhair tree	*Ginkgo biloba*	Ginkgoaceae
N, I	Maple	*Acer campestre*	Aceraceae
I	Monkey puzzle	*Araucaria araucena*	Pinaceae
I	Monterey cypress	*Cupressus macrocarpa*	Pinaceae
I	Mulberries	*Morus alba* *Morus nigra*	Moraceae
I	Nootka cypress	*Chamaecyparis nootkatensis*	Pinaceae
N	Oaks	*Quercus robur* *Quercus petraea*	Fagaceae
I	Pines	*Pinus*, many species	Pinaceae
I	Pittosporum	*Pittosporum tenuifolium*	Pittosporaceae
H	Plane	*Platanus acerifolia*	Platanaceae
N, I	Poplar	*Populus*, several species	Salicaceae
I	Redwood	*Sequoia sempervirens*	Pinaceae
N	Scots pine	*Pinus sylvestris*	Pinaceae
I	Southern beech	*Nothofagus antarctica*	Fagaceae
I	Spruce	*Picea excelsa*	Pinaceae
I	Swamp cypress	*Taxodium distichum*	Pinaceae
Nt	Sweet chestnut	*Castanea sativa*	Fagaceae
I	Sycamore	*Acer pseudoplatanus*	Aceraceae
I	Tulip tree	*Liriodendron tulipifera*	Magnoliaceae
I	Walnut	*Juglans regia*	Juglandaceae
I	Wellingtonia	*Sequoiadendron giganteum*	Pinaceae
N, I	Willows	*Salix alba* *Salix alba* *Salix aurita* *Salix capea*	Salicaceae
N	Yew	*Salix coerulea* *Taxus baccata*	Taxaceae

Chapter Three
Shrubs

THE shrubby flora of our island has been even more completely transformed from its primeval to its present aspect than the tree flora.

I believe that the best way to give some idea of this transformation will be to visit some of the great shrub gardens – there are several in every part of Great Britain – look at the plants growing there, and decide which are native and which are not. We shall make two such excursions: one in May and one in September. But one thing before we start; it would require volumes if we tried to deal with every genus; we shall confine ourselves to the more noticeable ones, the bushes we are more or less familiar with.

We shall begin in one of the great gardens of the extreme southwest, where many kinds of shrubs and small trees too tender to grow in the north or east flourish in the warm, humid climate. Entering the garden we see a group of small trees with light, feathery foliage and, if the month were January instead of May, flowers consisting of thousands of little, fluffy yellow balls with a delicious scent: **mimosa**. But the name is misleading, for this plant is a true *Acacia*, called wattle by the Australians. It was introduced from Tasmania in 1820: so we begin with a foreigner and a relative newcomer. By the way, and very confusingly, the tree we normally call **acacia** is not an *Acacia* but a *Robinia*, an American native called locust tree in its own country.

We are now walking along a broad grass path with shrub borders backed by tall trees on both sides. On the north side growing in shade are shiny-leaved evergreen bushes with masses of exquisite white or red or pink flowers – **camellias**. You do not have to come to the southwest to see them, they grow in thousands of gardens all over England, Wales and Scotland and they do well in city gardens; they are extremely hardy and will stand very hard frosts. But far from being natural here, nobody had seen one in Britain until 1720. They came from China and Japan.

On the south and sunny side of the path are massed tall, small-leaved bushes covered with lavender-blue flowers. They are called **California lilac**, though they are not lilacs at all. The first species ever planted in England arrived in 1713; later came a score of other species and they are now quite common in all but the coldest parts of England.

Turning off the grass walk on to a gravel path beside a wide border in full sun, backed by a wall, the scene changes completely. The border is full of low, dense shrubs covered with white and maroon, white and

Camellia, *Camellia japonica*

Common rock rose, *Helianthemum nummularium*

Forsythia, *Forsythia intermedia*

yellow, and pink flowers like very fragile single roses. And on the wall are some shrubs with narrow, spear-shaped leaves, hard to identify because they are not in flower. This shrub, trained to cover the wall, is **winter sweet**: it came here from China two hundred years ago and now there are many thousands in England. And the bushes in front of them are **rock roses** (*Cistus*), not one of them native, all of them introduced from south Europe since the seventeenth century. (There are four native species of the closely related genus *Helianthemum*.)

There is a big group of **forsythia** at the end of this path. The flowers are long since over – what would early spring in England be without the golden flowers, glowing with light, of forsythia? – but they are easily recognized and so common in millions of gardens that surely they must be native? Well, no: the first species of this genus to be introduced into England from China arrived in 1844; later came another from Japan.

Here, separating one side of the garden from the road, is a long, tall hedge of **fuchsia**. There are many such hedges in Cornwall and Devon and Wales, and even as far east as Suffolk; but the plant is no native here, it comes from southern Peru and Chile and was first tried in Britain only 150 years ago. Below the fuchsia hedge is a dense mass of very low, dark green shrubs with white flowers rather like the bells of lily of the valley: **creeping wintergreen** and other species of its kind – some are American and some from New Zealand, the first of them reached us in 1762 and none is native.

A great bed of **hydrangeas**, as common as roses in Britain, or almost. The most familiar species is *Hydrangea hortensis*; it was introduced, by Sir Joseph Banks by way of Kew, in 1789, from China where the species had already been transformed into a garden plant by many centuries of cultivation. These are the commonest kind of hydrangea with big spherical heads of flowers. The so-called 'lace-caps' were not introduced until 1879, this time from Japan. Meanwhile other species had come from the Himalaya mountains and from the south-eastern United States. The very useful climbing hydrangea, which clings like ivy and can attain thirty or forty feet, was introduced from Japan in 1878 and it is surprising that it is not more commonly planted here.

Next we come upon a group of small shrubs with long, narrow dark leaves, reminding us of rhododendrons, and bell-shaped flowers in white or pink. Recognize them? No – well, the Americans call them **sheep laurel** or **calico bush** and they were first planted here in 1736.

This garden has a 'wild' part, a dell with a stream running through it planted with **rhododendrons**. They flourish marvellously, but none is native here; and to explain them I will have to use their Latin names because there are no English ones.

The genus *Rhododendron* flourishes in Britain especially in the peaty soils of Scotland, and is more at home here than in any of the other countries to which it has been introduced. There are probably not far short of 900 species varying from sub-shrubs a few inches tall and with tiny leaves to eighty foot forest trees and huge bushes with leaves two feet long and a foot wide. One species, *R. ponticum*, has become a weed of some of our woodlands and is naturalized here: it has dark leaves and mauve flowers, is native to Spain, Portugal and Armenia (the Pontus of the ancients). It was introduced in 1763, before which date I doubt whether any rhododendron had been seen here. But this is not certain; there are two European alpine species, R. *hirsutum* and R. *ferrugineum*, for which Bean* gives 1656 and 1752 respectively as dates of introduction: they remained and still are rare in cultivation.

It would require a very fat book to tell the whole story of rhododendron introduction. A small number followed R. *ponticum* in the eighteenth century: the deciduous R. *canadense* from North America in 1767; R. *dauricum* from Siberia in 1780; R. *minus* from America in 1786; R. *camtshaticum* from north-east Asia in 1799; and some others. In the first quarter of the nineteenth century the pace quickened: the Himalayan rhododendrons – R. *anthopogon*, R. *arboreum*, R. *campanulatum* and many more – began to arrive, and more species, including some in the **azalea** section from America. In mid century, Sir Joseph Hooker introduced about forty species from north India. By 1900 the Chinese rhododendrons began to reach us; and so it has gone on until there are gardens with as many as 600 species and heaven knows how many hybrids and forms.

Up to now we have not seen a single shrub species native to England; they are all foreigners. We shall go further east to another garden and keep a look out, on the way, for bushy plants which *are* primeval natives and probably see some which are not.

Two great **barberry** (*Berberis*) bushes, with leaves like tiny holly leaves and masses of orange flowers, stand beside a gate; but this is Darwin's barberry from South America, introduced in 1849. You can find anything up to thirty species of barberry growing here in gardens, all foreign and all introduced in the last hundred years. It is true we have one native one, the common barberry, but you are unlikely to see it outside a nursery or a botanical garden. Its long racemes of yellow flowers and its bright red berries were, when the natural scene was set for us, very much more often to be seen than they are today. For one thing, progressive extension of cultivation has destroyed most of its habitat; for another, a deliberate attempt to exterminate it was undertaken late in the nineteenth century when it was discovered that the common barberry (but oddly enough none of its alien relations) is the over-wintering host-plant of the parasitic fungus *Puccinia graminis*, the black rust of wheat.

We are climbing on a long hill and emerging on to a small moor with a thin, peaty soil and great outcrops of grey granite, and a sea of the kind of **heather** called ling, which will turn the moor purplish-red with flower in August. There is no doubt about this being a native: moreover it is by far the most abundant of all native shrubs. Where it is dominant in the wild it creates an aspect which cannot have changed for thousands of years.

Not only that, but there are quite a number of native shrubs closely related to it: there is a heath (*Erica ciliaris*) found in Cornwall and Devon; there is the Scots heath or bell heather of Scotland and the West Country; there are also the cross-leaved heath, the Cornish heath and others. All these have been part of the English scene wherever the

* Bean, W. J., *Trees and Shrubs Hardy in the British Isles*, J. Murray.

Ciliate heath, *Erica ciliaris*

Overleaf: a flourishing shrub garden including two kinds of ivy, honeysuckle, lavender, rosemary, thyme and jasmine

41

soil is acid since time immemorial; they will not grow on the sweet limestone or chalk soils.

Coming down from the moor we run through woodlands where the rhododendron has gone native. If we had time to search those woods we might find another pretty shrub growing among the trees, the **Oregon grape**.

There is a small number of shrubs which, having been introduced from foreign parts, have become so thoroughly naturalized that the botanists have decided to list them, in future, as natives. For, after all, what is a native plant? We are using a special definition of the phrase 'natural to the English flora', but in fact, if a species flourishes in our woods, the manner and time of introduction are not significant. Among these special cases is the Oregon grape. This very attractive shrub was introduced in 1823 from western Canada, and nurserymen were charging £10 (about £100 in today's debased money) per plant. But propagation is so easy that by 1914 the price was thirty shillings (£1·50) per thousand. It is now not only a favourite garden evergreen, but is well-established as a wild plant in certain woods. Its close relation, *Mahonia japonica* (it has no English name – see p. 56), with scented flowers in midwinter, reached us in the mid nineteenth century from China and other, still finer, winter-flowering species still later.

Leycesteria (sometimes called Elisha's tears or Himalayan honeysuckle) is a curious case; birds, especially pheasants, are so fond of its very colourful berries that it was at one time widely planted as covert for pheasants. Its seeds germinate freely so that it readily naturalizes itself, and it is difficult to believe that this was no native but a happy immigrant from Himalayan forests, which first landed here in 1824.

Approaching the outskirts of a town, we see that although the ornamental almond-trees are 'over', some of the other members of the genus are still in flower: **plums**, **almonds**, **cherries**, **peaches** all belong to the same genus, *Prunus*, the one which also includes some of our native trees noticed in the last chapter – gean, for instance, bird cherry, and blackthorn. But virtually all the familiar ornamental shrubs and small trees of this genus are foreign immigrants: there is one, the **cherry laurel**, which does not look like a *Prunus* at all, but like a laurel, for it is evergreen. Planted by the million here in the eighteenth and nineteenth century, it is difficult to imagine England without it, but it was not introduced (from west Asia) until 1629, to be followed twenty years later by the Portugal laurel. Then there is the host of Japanese flowering cherries, all newcomers to our scene. The almond has been here longer, since the sixteenth century. The peach is found wild only in China; the ancient Persians had it from China about two thousand years ago and the ancient Greeks had it from Persia; and when they founded colonies in Italy, they took peach trees with them. The Romans introduced the tree into France and finally into England, so it has been here for quite a long time.

As we move further east there are fewer unfamiliar shrubs and more of the kinds most of us are used to. On our way there is a garden – in Dorset – with a wonderful collection of **roses** of all kinds. It is too early in the year to see most of them in flower, though there are a few species

Almond, *Prunus dulcis*

Cherry, *Prunus avium*

which flower in May; but we might as well pay it a visit.

Two genera of shrubs have, during the past three centuries, totally transformed the English floral scene: rhododendrons – with six or seven hundred species and hundreds of hybrids – now flourish here; and roses. We have five species native here but, except in the hedgerows, they are of no importance. In the rest of the world there are about a hundred species; and from a few of them have been made all the garden roses whose population is counted here in tens of millions. The story of those roses belongs to a later chapter. What is interesting about this garden is that its owners grow a lot of 'species', that is wild, roses from other parts of the world.

The huge rose plant, already covered with clusters of tiny, yellow, scented roses, covering almost the entire front of the house, is the Banksian rose; it is a Chinese plant and was introduced in 1824. Growing beside and over the front door, which faces south, is what is obviously a rose but equally obviously an evergreen; there are not many evergreen roses so it is not hard to identify as the Macartney rose; there is quite an amusing story attached to it.

In 1791 King George the Third sent an embassy led by Lord Macartney, carrying very valuable presents, to the Emperor of China. The idea was to establish trade with China. The Chinese of that time considered that they were the only civilized people in the world; that the world belonged to the Chinese empire; and that all countries beyond their frontiers were barbarian provinces of the Chinese empire so backward – from the Chinese point of view – that they were not worth bothering with. So Lord Macartney was received by the Emperor as a humble envoy come to pay his respects; his presents were received not as gifts from one great ruler to another, but as tribute from a petty chief to his overlord; and he was given a letter to King George in which the King was given a pat on the head for ruling his 'province' properly and told to carry on with the good work. Fortunately Lord Macartney had a gardener, charged with the duty of collecting new plants, with him: about all the British got out of that embassy was the Macartney rose.

There are about fifty species and several hundred varieties of roses from Russia, Europe, China, Japan, Mongolia and North America, all introduced to England in the last 200 years.

On some of the banks beside our road there are groups of golden-yellow **broom**: this is a primeval native but it is the only one of the genus. Thirty alien species have been introduced from Spain, Portugal, the Alps, Morocco, Greece, Hungary, the Canary Islands, and other places, giving us white, purple and multi-coloured brooms more spectacular than the native one. The same banks are often overgrown with trailing, sprawling vines of **traveller's joy**, our only native *Clematis*. Dozens of species have been introduced from all over the world, all of them with larger and more colourful flowers than our own.

An **apple** orchard is the next feature in the landscape, prompting the question – is the apple tree native? I have twice found what seemed to be wild apple trees growing in open woodland; I believe it probably is native, as it certainly is over a vast region of continental Europe. But

Banksian rose, *Rosa banksiae*

Traveller's joy, *Clematis vitalba*

Shrubs in bloom. *Top left: Chaenomeles; top right: Berberis darwinii; above: Hamamelis mollis; right: Camellia japonica; below:* 'Hortensia' hydrangea (*Hydrangea macrophylla*); rhododendrons of many colours.

Hedgerow fruit. *Above left:* guelder rose (*Viburnum opulus*); *above:* wayfaring tree (*Viburnum lantana*); *left:* sea buckthorn (*Hippophae rhamnoides*); *below left:* rowan (also called mountain ash, *Sorbus aucuparia*); *below middle:* elder (*Sambucus nigra*); *below:* blackberry (*Rubus fruticosus*)

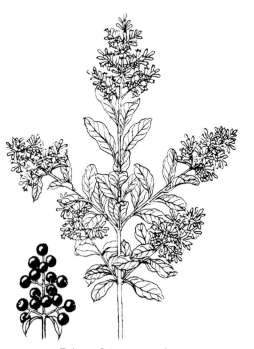

Broom, *Cytisus scoparius*

Privet, *Ligustrum vulgare*

that does not mean that our orchard and garden apples were developed here in England from the native wild one. The creation of orchard fruit trees by improvement of the wild ones in cultivation, by cross-breeding and selection, is one of the works of civilization and civilization in south-west Asia and south-east Europe was thousands of years ahead of civilization in Britain. Cultivated apple varieties were introductions. The same is true of other fruit trees. You can find wild **pear**, wild **plum** and even wild **medlar** in English woods, though I have never found the medlar. I have found all of them in Iranian woodlands and Iran is probably the land of their original 'domestication'. Yet I think they are among our primeval plants.

Now a coppice of **hazel** (see p. 54): this is certainly an ancient native, perfectly 'natural' to the scene. Yet, again, it is not the ancestor of our cultivated cob-nuts and filberts which came to us, via France–Italy–Greece, from south-west Asia, where they were produced thousands of years ago after centuries of selection.

We pass through a village: a small terrace of cottages is separated from the road by a **privet** hedge. This is a native English plant and, when allowed to grow naturally, a graceful one, though its pungent scent is not to everyone's liking. But although privet hedges used to be made with our native privet, they are now made from a Japanese species introduced in the last century, which has better foliage and is more reliably evergreen.

There is an obviously Victorian vicarage on the outskirts of the village and in its front garden a clump of **spotted laurel**. This plant had not been seen in England until 1783 when a Mr John Graeffer brought a single specimen from Japan. It is a unisexual species, which means that each bush bears either male or female flowers but not both. Species of this kind only produce fruits on the female plants, of course. Mr Graeffer's spotted laurel happened to be a female plant, but as there were no male plants in England and the females cannot produce fruits without pollen from the males, the original plant and all the young ones propagated from it by cuttings were barren. Then, in 1843 the famous plant hunter Robert Fortune introduced male plants of this 'laurel' from Japan, whereupon the females began to produce fruits; the fine oval, scarlet berries were a great improvement. Millions of these laurels were planted in Victorian gardens.

By the way, that word 'laurel' in 'spotted laurel', 'cherry laurel', 'Portugal laurel', etc., is misleading. Not one of them is really a laurel. The true laurel is **bay laurel**, *Laurus nobilis*, whose aromatic leaves are used in cooking: this is another foreigner, introduced from the South of France in the sixteenth century.

Just beyond the village, and before the real open country starts again, is a clump of a very familiar and very common shrub – **buddleia**, sometimes called butterfly bush because the butterflies are so fond of its flowers. Every English man or woman old enough to remember the late 1940s and 1950s will know that thousands of the bomb sites which still existed in the south of England were rapidly colonized by plants; wind-borne and bird-borne seeds sprang up and turned the sites into little jungles of weeds. Now one might have expected only truly native plants would have adapted to such colonization of brick and rubble. And most people assumed that the commonest of the shrubs which sprang up in such places – buddleia – must be native. But in fact the species is a Chinese plant not even discovered by a European botanist (Augustine Henry) until 1887; and the plant now 'native' to our waste places was introduced by way of France less than a century ago.

We have arrived at our second garden. Just inside the gate, in the shade of some trees, is a planting of the spurge laurel, *Daphne laureola* – yet another of those 'laurels' which are not laurels; and in front of them, in the sun, *Daphne mezereum*, now already covered with green berries, for it flowers in March. Both are ancient English natives. But further into the garden there are beds of shrubs which include many other daphnes, notably the lovely garland flower, all of them introduced within the last century.

There are **dogwoods** or **cornels** here in great variety, grouped round a big **handkerchief tree**, so-called because the bracts round the flowers look like handkerchiefs hung on the tree to dry (see p. 54). The story of its introduction to Britain is given on p. 114. These dogwoods, some evergreen, some deciduous, are from North America or from China. We do have a native cornel but it is not common in gardens because the foreign ones are more ornamental.

A wall overgrown with gold and silver variegated **ivies**: no plant is more natural in the English scene and all these colourful varieties have derived from the true native. So is the **honeysuckle** growing over an arch made of trellis, although as the finest varieties of cultivation were developed in Holland or Belgium, these varieties are called 'Dutch' honeysuckle.

Now comes a shrubby **cinquefoil**, *Potentilla*, with white, yellow and orange flowers. This is one of the few truly native shrubs – it can be found wild in the north of England – to provide us with a good garden plant. Is not **lavender**, here used as an edging plant, another? No: the so-called English lavender was introduced from the South of France in the thirteenth or fourteenth century, probably by monks for their monastery herb gardens. Another species of lavender came from the same source later.

While on the subject of herbs, we can take a look at the shrubby ones in the herb border of this garden. Here is **rosemary**; it always seems at home here, as happy as any native, but you will never find it outside a garden, and when we get a really hard winter it reveals its alien origin by failing to survive the worst frosts. It was introduced at about the same time as lavender from the coastal mountains of the Mediterranean. So was this handsome silvery shrub, called **lavender cotton**, which you will find in thousands of gardens but never in the wild. The tall-growing, woody-stemmed **thyme** used for cooking is another Mediterranean plant, though the pretty little creeping thyme, common on downland and moors, is English. This feathery green shrub with the strong spicy smell is **southernwood**, though some call it lad's love; it is a sixteenth-century introduction.

The **lilacs** are in flower in this garden; no flowering shrub seems more natural in the English scene. But it is nothing of the kind. There

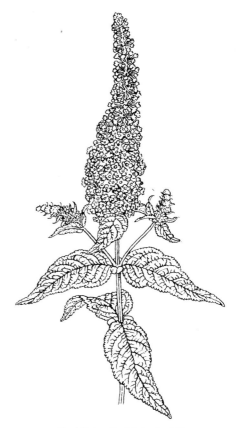

Buddleia, *Buddleia davidii*

Honeysuckle, *Lonicera periclymenum*

49

Laburnum flowers, *Laburnum anagyroides*

Virginia creeper, *Parthenocissus tricuspidata*

is a scene in the film *A Man for All Seasons* where King Henry VIII and Sir Thomas More compare the newly introduced lilac shrubs growing in their gardens. This they could not, in fact, have done, because no lilac had been seen in England until about 1600 when the first was introduced from France; the French got it from eastern Europe where it is native. Then, in 1640, we acquired a much better kind from Persia where it had been cultivated for many centuries. Since then others have been introduced from China and Japan.

Shrubby **veronicas** can be seen in the next border. They are familiar enough: you even find hedges of them beside the sea on the south and west coasts. But they come from New Zealand and Tasmania and have been here less than a century.

From here we can see the house and one wall of it is covered with a 'creeper' – the familiar **virginia creeper** – whose foliage, now pale green, will be flaming scarlet in the autumn. We tend to take it for granted since it covers the walls of tens of thousands of houses in town and suburbs and country. The first time I went to America in the autumn and saw these creepers growing over rocks and banks and up trees, I had the curious feeling that someone must have planted them. This species was introduced to England in 1629.

We come now to a long pergola tunnel covering thirty yards of a broad brick path, overgrown with two other climbing plants, **wistaria** and **laburnum**, trained to form this elongated floral archway, their pale mauve and deep yellow racemes of flower mingling above our heads. Neither is native: the first wistaria to reach England was an American species which arrived in 1724. But this you never see now because a much more beautiful one arrived from China in 1816, followed by another beauty, in 1830, from Japan.

As for the laburnum, whether grown as a tree, a shrub or, as it is here, trained over an archway, it has long been one of the half dozen most conspicuously ornamental trees in England and so common that I suppose most people think of it as native. In fact it had not been seen in England before Elizabeth the First came to the throne. It is a native of Central Europe.

Now we see a group of **'syringa'** – mock orange – not quite in flower yet. Here is another of those name confusions: the species we call 'syringa' is actually *Philadelphus*; and *Syringa* is the genus we call lilac. It is no more native than lilac, anyway. The first to reach England, probably late in the fifteenth century came from south-east Europe. Gerard, the seventeenth-century herbalist, objected to the flowers being brought indoors because their scent was strong enough to wake him from sleep. Although this is the mock orange blossom which is now so familiar, it is not the best we have: species introduced from China, California, Florida, Mexico, Colorado, Japan and the Himalayas, and hybridized during the nineteenth century, chiefly by French nurserymen, have given us much better ones.

You will remember that **'acacia'** was another of those name confusions; and that the tree we call 'acacia' is actually a *Robinia*. The garden we are visiting has a number of them. They are American shrubs and trees: the common robinia is a big tree, of course, but the pink-flowered robinia, introduced in 1743 is rarely bigger than a shrub.

Like most big gardens this one has several species of **spiraea** and several of the shrubby **senecio** species (no English names) planted more for their silver leaves than their flowers. They are all foreigners and all comparative newcomers. Very closely related to the senecios are the **daisy bushes**; they are all from Australasia and none of them arrived here before the nineteenth century.

Here is a great boscage of shrubs with big, cup-shaped yellow flowers, very like our native **tutsan** or **St John's wort** but much larger and with much bigger flowers. One of this genus which is very common and often seems to be growing wild, is **rose of Sharon**. It is not a native; it was introduced from south-east Europe in the seventeenth century. The taller, bushy one came from Japan (1862) and the most beautiful of all from Malaya about a century ago.

The path with the bank of exotic St John's worts ends at a small summer-house overgrown with two kinds of **jasmine**: one, sweet-scented and white-flowered; the other, a winter plant, leafless and yellow-flowered. Although you will find jasmine mentioned in English books as far back as you can go, it is certainly not natural here. The nearest place where this plant is found truly wild is Iran but it has been common all over the Middle East and eastern Mediterranean for thousands of years. Perhaps a soldier brought seeds of it home from one of the Crusades; or the monks may have introduced it in the twelfth or thirteenth century, probably for the scent which is extracted from the flowers.

As for the winter-flowering jasmine, it was not introduced, from China, until 1844, by Robert Fortune (see Chapter Seven).

Growing behind the summer-house, and spreading its branches over its roof, is a curious tree you will not find at all natural here: it has a woody trunk, long, very slender green branches and twigs, thousands of yellow pea-flowers, and no leaves at all. This is the **Mount Etna broom**, which is found wild in only one place, the slopes of Mount Etna. It was introduced here in the nineteenth century.

Three species of shrubs in this garden are so common and flourishing that they could easily be taken for natives: one piece of evidence that they are not is that they have never been given common English names but are known by the names of their genera: **deutzia, diervilla** and **weigela**. Deutzias come from Japan and China and none had been seen here before 1822. Diervillas started coming from America in the eighteenth century. And weigelas much later, from China.

One of the special sights of this garden is its **strawberry tree**: well over twenty feet tall, with a spreading head, red stems, dark laurel-like leaves, racemes of lily-of-the-valley-like flowers and handsome, strawberry-like fruits, it is one of the most beautiful large shrubs. It has never been found wild anywhere in England, Wales or Scotland and its real home is the shores of the Mediterranean. But oddly enough it is found wild in south-west Ireland and I suppose it was introduced from there.

* * *

Our September trip will not be as long as the May one; what we are looking for, chiefly, is berries; but there will be some flowers as well.

Mock orange, *Philadelphus coronarius* 'Beauclerk'

Strawberry tree, *Arbutus unedo*

Cotoneaster, *Cotoneaster horizontalis*
(shown vertical)

Going out through the suburbs you notice quite a lot of gardens with small trees or shrubs, some with narrow, hanging leaves, some with small rounded leaves, others growing prostrate or against walls in a curious fishbone pattern, but all loaded with red, orange, or occasionally yellow berries. They are **cotoneasters**, and though there is one native species, all the really ornamental ones, common though they are in cultivation, are foreigners, mostly from the Far East and most of them introduced in the last century; they have never acquired a popular English name.

It is quite easy to confuse them, at first glance, with the **firethorns**, to be seen on the walls of so many suburban houses, surely the most wildly colourful of all berrying shrubs, a solid mass of brilliant, scarlet fruits hiding the bright green leaves. It was introduced from south Europe in 1629. There is only one kind of shrub to compete with it for spectacular beauty in late summer, but we are not likely to see one today, since we are too far from the sea; this is the **sea buckthorn**. It is a primeval native, and the contrast between its silver leaves and masses of orange berries is very striking.

Those hedges with very deep green, glossy leaves – you have seen miles of them in your time and hardly noticed them – are **euonymus** – no English name, but an alien introduction related to our native **spindleberry** tree (see p. 56).

Once we get clear of the houses, the hedges are bright with berries – **rose** hips, **hawthorn** fruits and **blackberries**; they are ancient natives, of course; incidentally, so are **raspberries**, and our own wild ones are the ancestors of the garden kinds.

The **elderberries** are ripe; this is a native shrub and at one time in the past elderberries were cultivated in orchards. So valuable was every part of the bush, even the pith in the hollow stalks, that the famous Dutch naturalist Boerhaave never passed one without raising his hat.

Every now and again we notice in the hedges a big shrub – it can be quite a large tree bearing clusters of red fruit among its big, downy leaves; this is the **whitebeam**, another shrub natural to our scene, as is its relative, the **rowan** or **mountain ash**. Another bush of the same family, the **service tree**, has been found growing wild only twice and was probably introduced in the sixteenth century, although there is a native very like it which is found only in the southern counties.

Here is a patch of rough wasteland made beautiful by **gorse**; this has been here since the beginning, a primeval member of the flora as familiar to prehistoric man as to ourselves.

We come across a cottage garden with two berry-laden small trees. One of the trees is a **wayfaring tree** – the one with black berries. The other is a **guelder rose** – with vivid red berries. Both are native; and related to a number of fragrant, winter-flowering introductions (*Viburnum*), one being the evergreen **laurustinus**. Also in the cottage garden is a group of rather stiff-looking shrubs with leaves shaped rather like holly leaves but dull instead of shiny, soft instead of hard: they are covered with white, red and blue flowers like hollyhock flowers.

These **shrub hollyhocks** are aliens: they are actually hardy members of the *Hibiscus* family, which we associate with the tropics. Because they

Hawthorn, *Crataegus monogyna*

52

were first introduced (about 400 years ago) from a Syrian source, botanists gave them the specific name *syriacus*. But in fact they come from China originally.

The next garden on our route has gone in for **'japonicas'** in a big way and the bushes are covered with green and gold 'quinces'. The first of these ornamental 'quinces' was introduced by Kew Gardens in 1796, from China. But the first to become popular and to be planted by the thousand was a species which comes from Japan and was introduced in 1869. As it came from Japan, botanists gave it the name *Cydonia* (quince) *japonica* (of Japan). Now there are dozens, scores, probably hundreds of species of plants which have the same specific name – *japonica*; but for some reason the public chose to use the name for these ornamental quinces. By the way, if you look in the index of scientific names at the end of this chapter you will find that japonicas are now called not *Cydonia* but *Chaenomeles*; botanists decided that they are sufficiently different from the true quinces to deserve a new genus to themselves.

*   *   *

Among the shrubs named in this chapter are only the most familiar of those which have been introduced into our flora. A great many more are making a place for themselves but are not and maybe never will be common: the lovely autumn-flowering **eucryphias** from South America and Australasia, the Chinese and Japanese **witch hazels** flowering in midwinter; the fragrant, white-flowered **myrtles** and **osmanthuses**; the **pernettyas** from South America and the Chilean mountains with their strangely coloured berries – these and many others are enriching our cultivated flora and perhaps some of them will get a footing in the wild as other, earlier introductions have done, until the time comes when they are rated as native.

Gorse, *Ulex europaeus*

Rowan, *Sorbus aucuparia*

53

Hazel, *Corylus avellana*

Handkerchief tree, *Davidia involucrata*

ORIGIN		GENUS AND SPECIES	FAMILY
I	Acacia (locust)	*Robinia pseudoacacia*	Leguminosae
I	Almond	*Prunus dulcis*	Rosaceae
N	Apple	*Malus pumila*	Rosaceae
I	Azalea	*Rhododendron* spp	Ericaceae
I	Banksian rose	*Rosa banksiae*	Rosaceae
I, N	Barberries	*Berberis darwinii*	Berberidaceae
		Berberis vulgaris	
I	Bay laurel	*Laurus nobilis*	Lauraceae
N	Blackberry (bramble)	*Rubus fruticosus*	Rosaceae
N	Broom	*Cytisus scoparius*	Leguminosae
I	Buddleia	*Buddleia davidii*	Loganiaceae
I	Calico bush (sheep laurel)	*Kalmia* (several spp)	Ericaceae
I	California lilac	*Ceanothus* spp	Rhamnaceae
I	Camellia	*Camellia japonica* (and others)	Theaceae
I	Cherry	*Prunus* spp	Rosaceae
I	Cherry laurel	*Prunus laurocerasus*	Rosaceae
N	Cinquefoil, shrubby	*Potentilla fruticosa*	Rosaceae
N, I	Cornel (dogwood)	*Cornus mas* and other spp	Cornaceae
N, I	Cotoneaster	*Cotoneaster* (many spp)	Rosaceae
I	Creeping winter green	*Gaultheria procumbens*	Ericaceae
I	Daisy bush	*Olearia* (several spp)	Compositae
N	Daphne	*Daphne laureola*	Thymelaeceae
		Daphne mezereum	
I	Deutzia	*Deutzia* (many spp)	Saxifragaceae
I	Diervilla	*Diervilla* (several spp)	Caprifoliaceae
N, I	Dogwood (cornel)	*Cornus mas* and other spp	Cornaceae
N	Elderberry	*Sambucus nigra*	Caprifoliaceae
I	Eucryphia	*Eucryphia* (several spp)	Eucryphiaceae
I	Euonymus	*Euonymus* spp	Celastraceae
I	Firethorn	*Pyracantha coccinea*	Rosaceae
I	Forsythia	*Forsythia viridissima*	Oleaceae
I	Fuchsia	*Fuchsia magellanica*	Onagraceae
N	Gorse	*Ulex europaeus*	Leguminosae
N	Guelder rose	*Viburnum opulus*	Caprifoliaceae
I	Handkerchief tree	*Davidia involucrata*	Cornaceae
N	Hawthorn	*Crataegus monogyna*	Rosaceae
N	Hazel	*Corylus avellana*	Betulaceae
N	Heaths	*Erica ciliaris*	Ericaceae
		Erica cinerea	
		Erica tetralix	
		Erica vagans	
I	Hollyhock (shrub)	*Hibiscus syriacus*	Malvaceae
N	Honeysuckle	*Lonicera periclymenum*	Caprifoliaceae
		Lonicera belgica	

54

A suburban garden in Norfolk with laburnum, lilac (purple and white) and privet hedges in gold and dark green

Himalayan honeysuckle, *Leycesteria formosa*

I	Hydrangea	*Hydrangea* spp	Hydrangaceae
N	Ivy	*Hedera helix*	Araliaceae
I	Japonica	*Chaenomeles* (many spp)	Rosaceae
I	Jasmine	*Jasminum officinale*	
		Jasminum nudiflorum	Oleaceae
I	Laburnum	*Laburnum anagyroides*	
		Laburnum alpinum	Leguminosae
I	Laurustinus	*Viburnum tinus*	Caprifoliaceae
I	Lavender	*Lavandula spica* ·	
		Lavandula stoechas	Labiatae
I	Lavender cotton	*Santolina* (two spp)	Compositae
Nt	Leycesteria	*Leycesteria formosa*	Caprifoliaceae
I	Lilac	*Syringa vulgaris*	
		Syringa persica	Oleaceae
N	Ling	*Calluna vulgaris*	Ericaceae
I	Locust (acacia)	*Robinia pseudoacacia*	Leguminosae
I	Macartney rose	*Rosa bracteata*	Rosaceae
I	Medlar	*Mespilus germanica*	Rosaceae
I	Mimosa (wattle)	*Acacia dealbata*	Leguminosae
I	Mock orange ('syringa')	*Philadelphus coronarius* and other spp	Saxifragaceae
N	Mountain ash	*Sorbus aucuparia*	Rosaceae
I	Mount Etna broom	*Genista aetnensis*	Leguminosae
I	Myrtle	*Myrtus communis*	Myrtaceae

Mahonia japonica

55

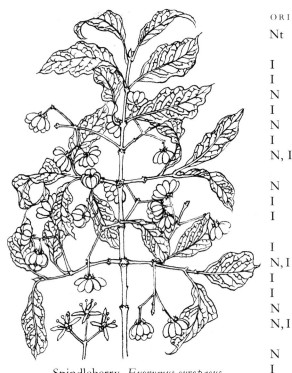

Spindleberry, *Euonymus europaeus*

Common laurel, *Prunus laurocerasus*

ORIGIN		GENUS AND SPECIES	FAMILY
Nt	Oregon grape	*Mahonia aquifolium* *Mahonia japonica*	Berberidaceae
I	Osmanthus	*Osmanthus* spp	
I	Peach	*Prunus persica*	Rosaceae
N	Pear	*Pyrus communis*	Rosaceae
I	Pernettya	*Pernettya* spp	Ericaceae
N	Plum	*Prunus domestica*	Rosaceae
I	Portugal laurel	*Prunus lusitanica*	Rosaceae
N, I	Privet	*Ligustrum vulgare* *Ligustrum ovalifolium*	Oleaceae
N	Raspberry	*Rubus idaeus*	Rosaceae
I	Rhododendron	*Rhododendron* (many spp)	Ericaceae
I	Robinia	*Robinia pseudoacacia* and *R. hispida*	Leguminosae
I	Rock rose	*Cistus* spp	Cistaceae
N, I	Rose	*Rosa* (many spp)	Rosaceae
I	Rose of Sharon	*Hypericum calycinum*	Guttiferae
I	Rosemary	*Rosmarinus officinalis*	Labiatae
N	Rowan	*Sorbus aucuparia*	Rosaceae
N, I	St John's wort (tutsan)	*Hypericum* (many spp)	Guttiferae
N	Sea buckthorn	*Hippophae rhamnoides*	Elaeagnaceae
I	Senecio	*Senecio* (many spp)	Compositae
N	Service tree	*Sorbus domestica*	Rosaceae
N	Service tree (wild)	*Sorbus terminalis*	Rosaceae
I	Sheep laurel (calico bush)	*Kalmia* (several spp)	Ericaceae
I	Southernwood	*Artemisia abrotanum*	Compositae
Nt, I	Spindleberry	*Euonymus* (many spp)	Celastraceae
I	Spiraea	*Spiraea* (many spp)	Rosaceae
I	Spotted laurel	*Aucuba japonica*	Cornaceae
I	Strawberry tree	*Arbutus unedo*	Ericaceae
I	Sweet bay	*Laurus nobilis*	Lauraceae
I	Syringa (mock orange)	*Philadelphus coronarius*	Saxifragaceae
N	Thyme	*Thymus serpyllum*	Labiatae
N	Traveller's joy	*Clematis vitalba*	Ranunculaceae
N, I	Tutsan (St John's wort)	*Hypericum* (many spp)	Guttiferae
I	Veronica, shrubby	*Hebe* (many spp)	Scrophulariaceae
I	Virginia creeper	*Parthenocissus* *quinquefolia*	Vitaceae
I	Wattle (mimosa)	*Acacia dealbata*	Leguminosae
N	Wayfaring tree	*Viburnum lantana*	Caprifoliaceae
I	Weigela	*Weigela* (several spp)	Caprifoliaceae
N	Whitebeam	*Sorbus aria*	Rosaceae
I	Winter sweet	*Chimonanthus praecox*	Calycanthaceae
I	Wistaria	*Wisteria frutescens*	
		Wisteria floribunda	Leguminosae
		Wisteria sinensis	
I	Witch hazel	*Hamamelis* (several spp)	Hamamelidaceae

Chapter Four
The Food Flora

BECAUSE England is and for so long has been an intensively cultivated land, its plant aspect is, except in a very few places, largely composed of what are called 'economic' plants – those grown to provide food or industrial materials such as timber, fibres for making fabrics, and so forth. When we look out over the English countryside it has a deceptively 'immemorial' aspect, despite the changes of recent decades. But in reality it has undergone great and continuous changes and in this chapter I propose to look at some of them.

A lovely feature of many English landscapes is the **apple** orchard, whether in flower, in fruit or even, in its regularity and the patterns formed by the naked branches against the sky, in midwinter. Just how old is this feature and how much has it changed?

Malus sylvestris, the – or an – ancestor of the cultivated apples, is native in England, all Europe, Anatolia, parts of Iran and the Caucasus. Although it has been established that apples were valued as food in prehistoric Europe, I think that the first orchards were planted in England in pre-Roman Celtic times, but the European continental Celts whose descendants colonized England probably had the idea of such plantations from their east European and west Asian ancestors, because by far the most ancient tradition of apple orchards is either German or Armenian. There is, at all events, a great deal of literary evidence that the west European Celts planted orchards. We can safely take it that orchards were first planted in England not later than 100 B.C. and have therefore been a part of the landscape for 2000 years.

But that does not mean that their aspect has not changed. A primitive orchard would have had none of the regularity of those we are used to: some trees would have been small, others large and there would have been differences of form and habit. In due course the progress of grafting techniques, of selection to establish the best varieties, and of methods of cultivation, would have introduced the order and regularity that we are used to. Later, came another change. The older orchards are composed of large trees: if you grow an apple tree from a pip it may be twenty years before you get any fruit. It was discovered that if you graft a desirable apple variety on to a root-stock which causes the tree to be dwarfed because the root is relatively inefficient, the tree, though much smaller in stature and bulk, will bear fruit much sooner – in three or four years. So orchards have been getting progressively lower in stature: an old orchard presents the aspect of a regularly planted forest; a new orchard is shrubby.

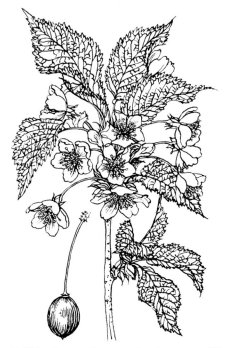

Edible cherry, *Prunus avium* (see page 60)

The wild crab apple, small and sour (*below left*), has been developed over the centuries into the many present-day varieties with far bigger and sweeter fruit (*adjoining drawing*). One of our outstanding eating apples is Cox's Orange Pippin, photographed in flower and fruit (*left*); it was raised well over a century ago.

Methods of cultivation were greatly altered by grafting the selected varieties on to special rootstocks. In the past, seedling rootstocks were used (*upper root drawing*), which usually grew into large, unmanageable trees (*bottom left*), slow to come into bearing and uneven in size. This situation was transformed by using uniformly grown rootstocks (*lower root drawing*), selected for consistent vigour whether strong or weak; hence the final size of the tree could be much smaller overall (*bottom centre*).

Several methods of propagation are used, including budding (*upper right centre*), in which buds sliced from the fruiting variety are inserted into T-shaped slots on the rootstock (the enlarged, newly cut slice is shown *in the centre*); and whip-and-tongue grafting (*below right*), in which a piece of one-year-old wood carrying buds is trimmed to fit a matching cut in the rootstock.

In all cases the piece of stock that remains above the new growth produced from the bud or graft-wood (*right*) is cut off the following spring.

Pear, *Pyrus communis*

Plum, *Prunus domestica*

Pear orchards are not as common in England as apple orchards but still there are a great many, and in some parts of the country they are important in the landscape – as are the remaining magnificent old specimen perry-pear trees in Herefordshire, surely the most beautiful flowering tree of forest-tree size. Pears formed a part of the diet of prehistoric Europeans – sliced and dried for winter use and accidentally preserved by carbonization, they have been found, like similarly treated apples, on Neolithic lake dwellers' sites in Switzerland and Italy. No doubt the fruit was gathered from wild trees; but cultivation of pears is probably older than civilization in England – meaning by civilization the beginning of urban life. It is significant, for example, that the vestiges of apples and pears on Bronze Age sites show improvement in the size of fruit over those found on Neolithic sites, for this implies cultivation and selection.

It seems, then, that some kinds of orchards were an element of the cultivated plant-scape more than 2000 years ago. But for centuries there were only apple and pear orchards. Even in Roman Italy **plum** orchards were not common until Julius Caesar's time – to Cato (234–149 B.C.) they were a rarity, to Virgil (70–19 B.C.) a commonplace. The Romans probably introduced plums, **peaches** and **cherries** to England but there is no evidence of orchards of them until the fifteenth century – perhaps such plantations were first made in the fourteenth century.

One of the pleasantest sights in Kent and a few other places in England, comparable in its almost geometrical order to a hop garden, but even more pleasing as a pattern, used to be the **filbert** and **cob-nut** orchards. The method of pruning the trees produced a shape which, reduced to an abstraction, was a wheel on a post. Perhaps there may be such a nut orchard left; if so, I do not know of it. I have dealt in Chapter Three with the introduction of these nut trees into our flora. They are unlikely to vanish from our flora entirely, but the decline of nut orchards has altered the aspect of parts of England in the last fifty years.

How long have there been great fields of **strawberries** in Kent, Hampshire, Dorset, Devonshire and East Anglia? Are they 'immemorial'? Very far from it: the strawberry we know is quite a modern fruit. The strawberries native to the Old World are the small and rather dry berries of *Fragaria vesca*, which was first taken into cultivation early in the fourteenth century; and the not much larger, musky-flavoured berries of *Fragaria elatior*, known in cultivation and in the England of the eighteenth century, as 'hautbois' strawberries. Strawberries of the kind we are familiar with – the strawberries of, as it were, strawberries-and-cream – simply did not exist until the nineteenth century and are 'man made'. When North America was explored the species *Fragaria virginiana* was discovered and introduced. The berry was larger than our native one but nothing like as large as the modern cultivated strawberries. Then in the eighteenth century the much larger-fruited *Fragaria chiloensis* was brought to France from the island of Chiloë off the coast of Chile: that was in 1780. The French botanist Duchesne crossed the two American species; one of the resulting seedlings had larger fruit and better flavour than any strawberry then in cultivation and this was

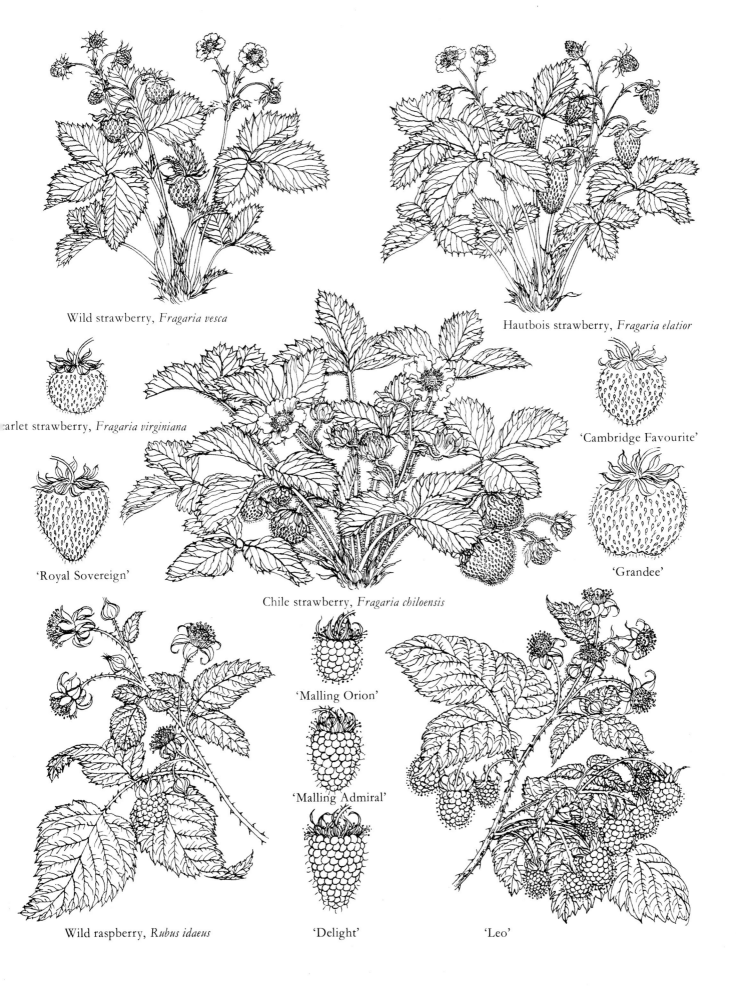

Wild strawberry, *Fragaria vesca*

Hautbois strawberry, *Fragaria elatior*

Scarlet strawberry, *Fragaria virginiana*

'Cambridge Favourite'

'Royal Sovereign'

'Grandee'

Chile strawberry, *Fragaria chiloensis*

'Malling Orion'

'Malling Admiral'

Wild raspberry, *Rubus idaeus*

'Delight'

'Leo'

Peach, *Prunus persica*

put on sale with the name of 'ananas' because it tasted of pineapple (*ananas* in French). An English nurseryman named Michael Keen started his own programme of breeding these with plants of *F. chiloensis* and *F. virginiana*. Finally, in 1821, 'Keen's Seedling', ancestor of the modern large-fruited, high-flavoured strawberry, came on the market.

It was, in short, not much over a hundred years ago that fields of strawberries as we know them first appeared on the English scene.

Much older in the garden flora are relatively large-fruited and luscious **raspberries**, and for an excellent reason: it is a native. Although it did not become commercially important till the nineteenth century, there were both red- and white-fruited kinds in cultivation by the sixteenth. It is probable that this fruit was taken into cultivation by cottagers very early, perhaps in the thirteenth century, and was steadily improved by selection and care for centuries before nurserymen began to take an interest in it.

* * *

If you had taken a walk through Kent or Norfolk or Gloucestershire at any time between the tenth and late fourteenth century, you would have taken for granted a feature of the cultivated landscape which was to vanish – though slowly – between the fourteenth and seventeenth centuries and which is now quite possibly to be restored to us again: **vineyards**. But suppose you had not been born in the fourteenth century but in the twentieth and had been given the privilege of going back six hundred years to take a walk through the Kentish countryside, you would be surprised to see that there were then no **hop** gardens there. So 'typical' of Kent and Herefordshire is the beautiful, regular pattern of the hop garden that it is difficult to imagine either county without them – and, by the way, without the oasts which go with them.

Right: 'Harvesting the Grapes' from *The Kalender of Shepherds*, 1493
Above right: Hop harvesting in Kent in the 1950s. Today virtually all picking is done by machine

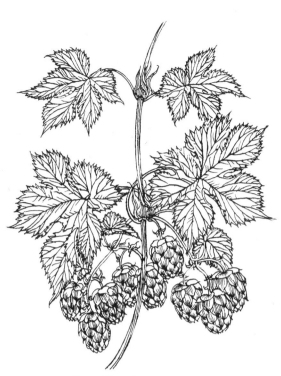

Hop, *Humulus lupulus*

The hop – *Humulus lupulus* – is native to England, but as a cultivated plant whose flowers were used to give flavour to beer, it was imported from Holland late in the fifteenth century. So the 'immemorial' hop fields of Kent and Hereford are, in reality, only five centuries old at most. As the vineyards disappeared, so the hop gardens appeared. Why? For one thing, a cycle of harsher weather began about 1350 and pushed the vineyards southwards all over Europe; for another, in the Plantagenet Franco-English empire, wine could be so cheaply produced in Gascony that the English growers were squeezed out by a cheap – and better – imported wine.

There is an old rhyme which goes:

> Turkeys, heresy, hops and beer
> Came into England all in one year.

I doubt whether it is literally true; I first came across it quoted by Rudyard Kipling in an enchanting story called 'Hal o' the Draft', but it was years and many rereadings later before I noticed a curious mistake in the story: Hal, its craftsman hero from the fifteenth century, is familiar with the growing habit of **'runner' beans** but has never seen hops – or turkeys. Now, both runner beans and turkeys are a product of Mexico, 'domesticated' creatures of the Maya–Toltec–Olmec–Aztec civilization. And Hernán Cortés did not complete the conquest of Mexico until 1520. A man who had never seen hops in cultivation could not have been familiar with runner beans either; and it is unlikely that a man who was familiar with runner beans could have been amazed at the sight of a turkey. Not that it matters: the story is still enchanting.

*　　*　　*

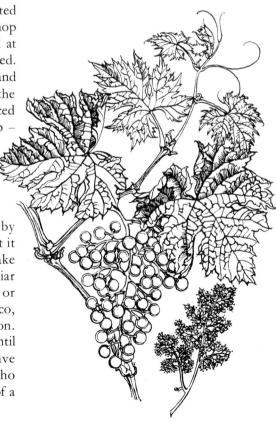

Grape vine, *Vitis vinifera*

Now let us go back much further in time. Can you imagine an English plant-scape without fields of **wheat**, **barley** or **oats**? It is not surprising if you cannot because they all go back a very long way in time. But they are not 'immemorial' and it would be an exaggeration to claim that they go back 'till the memory of man runneth not to the contrary'.

Men, before someone thought of farming, ate the seeds of some grasses just as birds do. They began deliberately planting them, first barley, then wheat, oats much later, in what is now Israel, and it took a very, very long time for the art to reach England. The first 'farming' people here exploited grasses in a different way, using sheep, cattle and pigs to convert them to meat.

Arable farming based on cereals as a way of life began in Asia Minor before 5000 B.C. The new art of wheat and barley growing slowly spread outwards from the original centres; villages based economically on wheat or barley growing were probably in being in England by the fourth millennium B.C. At all events, wheat and barley fields are so ancient in the landscape of cultivation – although they could not have looked like our own vast fields – that their absence from the scene would not have been noted by a time-traveller until he was back into the fifth millennium B.C.

You will note that I have not mentioned oats: the reason is that they are latecomers to our scene. The other grains we had from the east by way of continental Europe. But although the Mysians of Asia were growing oats as fodder for their horses early in the second century A.D., neither the ancient Greeks nor the Romans grew oats, so we cannot have had this plant from the usual sources of our culture. So where did this crop come from, and when? The elder Pliny, writing between the years 43 and 79 A.D. said that the Germans ate oatmeal. I think what happened was this: when the art of agriculture was brought to north-east Europe or north-west Asia by migrants from the original centres, they found that wheat and barley were apt to let them down by failing to ripen in the short, cool northern summer. (Now, of course, we have strains of these cereals especially bred for our climate.) They therefore took into cultivation the best wild cereal they found growing in their own part of the world, that is to say the oats they were already accustomed to gathering in the wild. The practice of cultivating oats then spread all over north Europe. I imagine that some time between 200 and 300 A.D., there was something new in the food-plant scene: patches or small fields of oats. It is even possible, however, that this crop did not arrive until a century or so later: with the Teutonic colonists and conquerors of England.

We are, nowadays, as familiar with fields of silk-tasselled **maize** on our farms, and **sweet corn** in our gardens, as we are with wheat and barley and oats. Of the cereals it is the latest arrival. The plant is a native of South or Central America; it was developed as a food crop by the peoples of the Andean and Central American cultures and it was the economic foundation of the great Andean cultures which were united by the Incas into the huge empire destroyed by Pizarro and his Spaniards; and of the Central American urban civilization under the imperial dominion of the Emperor Montezuma, which Cortés conquered

Indian hemp, *Cannabis sativa* (see page 72)

64

in 1520. Just how soon after Columbus reached America seeds of maize arrived in Spain, we do not know, but, obviously, not much before 1500. In an incredibly short time this magnificent grain had spread east across Europe and Asia into China and south into Africa. It reached England in the sixteenth century, but it was not for another two centuries that maize fields were to be seen in the landscape.

*   *   *

The kitchen gardens of our private houses, the vegetable-growing allotments and the great areas – Thanet for example and parts of Cornwall – given over to market-garden crops, compose another element of the plant population which we should miss if it were not there, finding England not quite England without them. Yet they are relatively new upon the scene. The **cabbage** patch or kale-yard is the most ancient element of the kitchen and market gardens.

The cabbage is a native of maritime Europe including Britain. There is good·evidence that the cabbage was first taken from the wild and cultivated in Western Europe. But when? The Greeks, though it is not found wild in Greece (it is in Anatolia), were familiar with it in the third century B.C. There could, of course, have been more than one 'domestication'. The Romans had six varieties by the first century A.D. and probably introduced it to England a century later, if the south British had not long since had it from Gaul. At all events it is safe to say that there have been cabbage patches in England for eighteen centuries, possibly rather more.

What, then, about cauliflowers, broccoli and brussels sprouts, all so much a part of our scene today. Are they, also, two thousand or so years old in the English plant-scape? By no means. We shall take the **cauliflowers** first: Pliny (23–79 A.D.) describes a vegetable which sounds like **sprouting broccoli**, which he calls 'Cypriot cabbage' and which presumably came from Cyprus. But the earliest description of what was unmistakably a cauliflower is in a twelfth-century Arabic treatise whose author, Ibn-al-Aram, calls it Syrian cabbage. Probably the cauliflower first appeared as a mutant cabbage in Syria, perhaps as early as the tenth century. We know that the Italians got their first cauliflower seeds from the Levant and were cultivating cauliflowers in the fourteenth century. I do not believe we had cauliflower fields before the fifteenth century and maybe not until the sixteenth century, and the vegetable did not become common until a good deal later.

As for **brussels sprouts**, they came in much later still. The odd thing about them is that they could have been with us much earlier and why they were not is a mystery. Here is their history. They appeared as a mutant of the 'Savoy' cabbage, probably in what is now Belgium, and probably in the twelfth century. Why Belgium and when exactly? Well, the earliest reference we have to them as *spruyten* is in the market regulations of certain Belgian towns very early in the thirteenth century. There are two other early references – under the name *sprocq*: one is in the accounts ledger of the Duke of Burgundy touching the food bought for

Wild cabbage, *Brassica oleracea*

Cauliflower, *Brassica oleracea* 'Botrytis'

Old-fashioned kitchen gardens like the one shown here are becoming rarer, though some can still be found in stately homes. The enclosing wall, with its trained espalier tree, helps to trap and concentrate heat for ripening fruit

A special feature of this garden is its
lean-to greenhouse, in which can be
seen a vine, tomato plants and a
peach tree

Brussels sprouts, *Brassica oleracea*
'Gemmifera'

Baudouin de Lannoy's wedding to Michielle Denne in 1472; and another in similar accounts for the wedding of Alcaude de Brederode in 1481. So, by the end of the fifteenth century the Belgians had been eating brussels sprouts for three hundred years. Yet brussels sprouts were not seen in France until the eighteenth century and made their first appearance on the English scene in the nineteenth.

Why? There is no satisfactory explanation. The English were always resistant to new vegetables and it is a fact that all the vegetable and salad crops were much later on the English than on the Continental scene. But why the French, with their passion for good food, should have waited five hundred years before introducing brussels sprouts into their diet, heaven knows.

What of our **spinach**? This is another relative newcomer. The probable route of its arrival is as follows. A native plant of Iran, *Spinacia tetrandra* was first cultivated there about A.D. 500. One of the Persian names for it was *aspanaj*. The Arab conquerors of Iran took a liking to the stuff and spread it to all the lands they conquered, including Spain, mispronouncing its Persian name, which thus become *isfandsh* in Arabic. Spinach was cultivated then, in Moorish Spain from about the ninth or tenth century, but just as brussels sprouts did not reach France or England from Belgium for centuries, so spinach did not reach Christian Spain – that is the north – for centuries. The Spaniards hispanicized the Arabic *isfandsh* into *espinaca*. We had our first spinach – we simply dropped the initial *e* and terminal *a* – from Spain during the reign of Elizabeth I. The first recorded use is in 1568.

In the year 1551 the conqueror of Inca Chile, Valdivia, wrote a report to the Emperor Charles V. In it he mentioned some of the crop plants cultivated there; among them was a root crop new to him: it was the **potato**. How long was it before the new food plant was heard of in England? Well, in 1571 John Frampton published an English translation of Doctor Nicholas Monardes' account of plant introductions into Europe from the Americas: the potato is not in the list. But it does appear in the 1596 edition of Gerard's *Herbal* and it is obvious that Gerard had actually grown potatoes. We can take 1590 as the date of its introduction into England, although it was known to some Englishmen before that date – Francis Drake took on potatoes as part of his ship's stores at the island of Mocha off the coast of Chile in 1577.

But the fact that there were a few potatoes growing in England in 1590 does not mean that they quickly became familiar plants in the English farm and garden flora. Far from it: only in Ireland did they quickly get a place – potatoes had become a staple food in parts of Ireland as early as 1610. In England the story was very different. Potatoes *were* grown in a few market gardens here as early as that but they retailed at an appallingly high price – one shilling a pound (a shilling in 1610 was worth at least £1 of our money). By mid century supplies were plentiful and by the third quarter of the century the price had fallen although 'it was still a luxury in the kitchen of Oliver Cromwell's wife, Joan'.*

It was in Lancashire that potatoes first became really familiar in the

Carrot, *Daucus carota*

* Eden, Sir F., *The Status of the Poor*, 1797.

scene, in cottagers' gardens first, and then in farmers' fields. But it is remarkable that Defoe, in his *Tour of England* (1774), does not even mention the potato. Evidently, then, it was not until the nineteenth century that the food plant developed by the ancient Chileans, who found it wild in the Andes mountains, became a really major element of our agricultural flora.

I made a passing reference to **carrots** in an earlier chapter. They are older in the farm and garden flora than potatoes; but not much older. Purple carrots were being grown in Dutch market-gardens in the fourteenth century; and also, but more rarely, white carrots. The Dutch had them from Christian Spain, which had had them from Moorish Spain, where they were cultivated at least as early as the eleventh century. The Moors in Spain had it from their co-religionists, probably in Afghanistan, where not only is the plant found growing wild, but where there are, or were, by far the greatest number of cultivated varieties, always good evidence for great antiquity in cultivation.

So it was from Holland we had our first carrots which were established in our market gardens during the fifteenth century – both the purple and the white. As far as we know, nobody had yet seen the orange-coloured carrots we are familiar with.

Purple carrots are, from time to time, apt to produce a yellow mutant. Several peoples, including the Dutch, picked out these yellow carrots as a separate variety.

Among the yellow carrots there appeared, in the seventeenth century in Holland, deep orange mutants. The Dutch market-gardeners selected and segregated them and produced a strain of orange carrots. These, too, we had from Holland. Why did white, purple and yellow carrots disappear and orange carrots – in no way superior – become *the* carrot? Simply because of the colour: they look a more appetizing colour than the others when cooked and so market pressure determined what the familiar carrot of commerce should be like. But it has not been familiar for more than two centuries.

There was a time in England when you could pay your rent in **onions**: that was in the early Middle Ages, so we know they have been a part of our garden and farm flora much longer than carrots or potatoes. The plant is *Allium cepa*; it was probably domesticated somewhere in the Iranian region of Asia; the Greeks had it in their kitchen gardens at least as early as 800 B.C. They had it from Crete and the Cretans had it from Egypt. When the historian Herodotus visited the Great Pyramid of Cheops at Gizeh as a tourist about twenty-five centuries ago, there was a plaque on the Pyramid recording the fact that in the course of its building 1600 talents of silver had been spent on onions, leeks and garlic for the building workers.

The Romans had onions from the Greeks and brought them to England (if they had not reached us even earlier) – so they are quite ancient here. If you consider the leek (*Allium porrum*) as a kind of onion, then onions are perhaps older still in our garden flora because the leek is a cultivated race of a native *Allium* – *A. ampeloprasum* – which is only found on Steep Holm in the Bristol Channel. Perhaps that is why it is the Welsh national plant. At all events, while most of our familiar vegetables

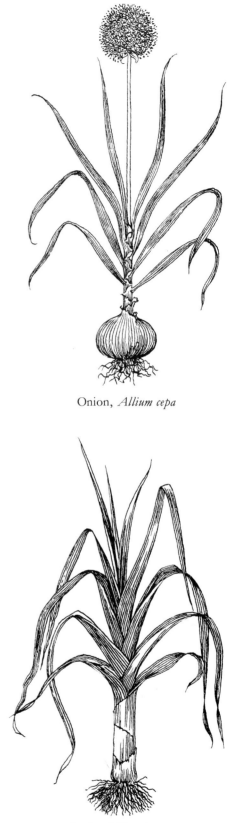

Onion, *Allium cepa*

Leek, *Allium porrum*

69

Turnip, *Brassica rapa* 'Sprinter'

Jerusalem artichoke, *Helianthus tuberosus*

Pea, *Pisum sativum*

are relative newcomers, the onion is ancient in our cultivated flora.

Let us look more briefly at a few other edible vegetables. **Turnips** and **swedes** are European, fleshy-rooted mutants of *Brassica rapa*, a close relation of the cabbage. Carbonized turnips have been found on Neolithic sites, so there is nothing new about them in our scene. Their nearest relative in our cultivation is a crop called rape, cultivated *Brassica rapa*, and it was probably in a field of this that a turnip made its first appearance as a mutant. Dating the event is impossible; it is more ancient than metal tools.

The **Jerusalem artichoke** does not come from Palestine and is not an artichoke; it is, in fact, a sunflower; the French for sunflower is *girasol* because the flowers turn on their stems so that they are always facing the sun; 'Jerusalem', in this context, is an English corruption of the word *girasol*. As for 'artichoke', the name was given to the vegetable when it was new because this root is supposed to taste like the much more ancient (in Europe) globe artichoke.

How did we get this plant, which flourishes so rampantly in England that even the natives have no chance against it? In 1534, Frenchmen had established a colony in the Bay of St Lawrence in Canada. When the great explorer Champlain, sent out by Henri Quatre, arrived to take charge of the colony, the people were starving. Champlain inquired into the diet of their neighbours, the Algonquin Indians, and in 1603 reported that they cultivated in their gardens a nourishing root crop which tasted like artichokes. As a result, it was introduced into France and England, where it might have become the most commonly eaten root vegetable but for the competition of that other American food plant, the potato.

How about **celery**? It is another newcomer. The plant is *Apium graveolens* and it is native to all Europe, including England, and is first mentioned in English writing in one of the Paston letters, from Italy; the writer says he has tasted a new vegetable called *celeri* but 'that it is nothing but the smallage'. But there is no evidence for its cultivation here before the sixteenth century. The Italians at first seem to have used it as we use parsley, chopping it up as a condiment. The practice of eating the blanched stems was new there in the fifteenth century.

Broad beans? The plant is a native of the Mediterranean basin and was in cultivation in Egypt as early as 2400 B.C. It is prehistoric in England and has been an element of the cultivated landscape since Neolithic times. Presumably it reached us in the luggage of an immigrant early farmer. The history of the **pea** is much the same except that they seem not to have been cultivated in ancient Egypt. Both are 'immemorial' in England.

The case of the **haricot beans**, **scarlet runners** and **French** or **kidney-beans** – in short, members of the genus *Phaseolus* – is very different. There are two species involved, both natives of Central America and of the Andes. They were originally domesticated by either the South American or Central American Indians. The first European to describe them was Christopher Columbus, who saw fields of these beans in Cuba three weeks after his arrival in America. They must have been in cultivation in one of the great centres of native American civilization for a very long time, for they were being grown by relatively backward tribes as far north as Canada. So the kidney-beans were all very new here in the late sixteenth century.

Lettuces were probably first grown here in Roman gardens, then lost and reintroduced in the early Middle Ages. They are older in garden and farm flora than most of our vegetables, then, but not among the most ancient.

Fields of **sugar-beet** are quite as familiar to all of us as fields of wheat or barley. But they have only been so for fifty years; you would have scoured all England in vain for a field of sugar-beet before 1914.

The plant is *Beta maritima* and it is also the parent of the salad beetroot and other garden and farm roots. It is native to the Mediterranean coasts and to southern England. Various kinds of beetroot developed from it had been in use as early as the fourteenth century; the Greeks and Romans knew the plant but in their day the fleshy root had not been enlarged by selection and they ate the leaves only, as a salad or green vegetable.

During the Napoleonic wars the British Royal Navy's blockade of the European ports cut Europe off from the source of cane-sugar – the West Indies. So French farmers set about developing the sugar-beet as a farm crop, continued the work during the nineteenth century and made France independent of foreign sources of sugar. England continued to import sugar until the First World War, and then introduced the beet from France in an effort to produce all its own sugar. The industry began to expand after the war; so that it was not until the 1920s that sugar-beet fields became a familiar sight in England.

The fruit with the greatest sale on the English market is not, as you may believe, the apple; it is the **tomato**.

Lycopersicon esculentum is a member of the family Solanaceae, to which also belong the potato, and such poisonous plants as deadly nightshade and henbane. It is a native of Ecuador and was domesticated and improved in cultivation over many centuries by the South American

Haricot bean, *Phaseolus vulgaris*

Below left: tomatoes growing under polythene
Below: a detail of a tomato plant showing multiple fruit on each plant

Flax, *Linum usitatissimum*

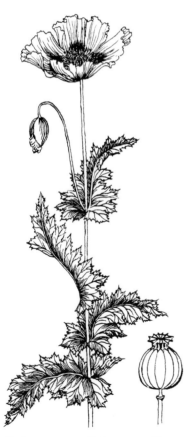

Opium poppy, *Papaver somniferum*

Indians. It was first seen by Europeans in Mexican gardens and seeds reached Spain about 1525. With it came the Aztec name for the fruit – *tumatl*. It got a bad start in Europe because Matthiolus of the Padua botanic garden, the first European scientist to examine the plant, condemned it as unwholesome – *mala insana*. The Italians, however, began cooking and eating tomatoes about 1560 while the French, English and Germans continued to regard them as poisonous and grew the plants as ornamental curiosities. This continued throughout the seventeenth and eighteenth centuries, but the practice of eating tomatoes was slowly creeping northward from Italy and in the nineteenth century the English began at last to treat the tomato as a crop plant to be taken seriously. Distrust of it lasted among old people into the twentieth century.

So the importance of the tomato in our farm and garden flora is less than a century old; and the overwhelming popularity of the fruit less than half a century old.

*   *   *

There are some plants which, having once been elements of our cultivated flora, have now virtually disappeared. For example, **flax** fields, with their pretty blue flowers, were common in England until the late nineteenth century. Now, as a consequence of the domination of the textile market by cotton and by man-made fibres, and the decline of linen, they are much less so. That was a big change after a very long history. The cultivated 'races' of flax are of two kinds: those bred specifically for fibre to make linen; and those bred to produce the oily seeds from which linseed oil is extracted and which make a valuable cattle-food. For how long has flax been with us? The question is unanswerable, for so ancient is the cultivation of this plant that it is very much older than civilization. We cannot even say when and where linen was first woven, for flax either has a range in nature so vast that it could first have been used in scores of different places; or it has become naturalized as an escaper from centres of cultivation in so many places as completely to obscure the evidence.

We know that the Egyptians had linen more than 6000 years ago. We know that Neolithic Europeans grew flax and used it to weave linen. All we can be sure of is that flax is at least as old as barley in our farm flora.

The only other plant grown in England for its fibres is **hemp**, and that has almost vanished from the scene. *Cannabis sativa* is notorious nowadays as the source of 'pot' and 'grass'. It is a Siberian native and was first cultivated by the Scythians as an intoxicant and for its fibres, from which rope was made. In India and some other eastern countries it was more valued as a drug plant than as an industrial plant. But in Europe it was grown for its fibre and as a fibre plant was introduced into our own farm flora some time in the early Middle Ages, probably from Russia via Germany. I doubt whether the Lincolnshire and East Anglian farmers even knew that hemp is a drug plant; and there is a good reason – the toxic attribute seems not to develop properly in our relatively cool summers (see drawing on page 64).

As a matter of fact no stimulant or narcotic drug plant has ever gained much footing in England. The **opium poppy**, *Papaver somniferum*

– a cultivar unknown in the truly wild state, probably a mutant in cultivation of the oil-seed poppy *Papaver setigera*, and naturalized all over Europe, especially as a weed of cultivation – grows readily in England but as far as I know has never been cultivated here as a crop. That is rather curious, really, because the opium poppy is a plant of Mediterranean origin, cultivated very early in Greece and Etruscan Italy. Despite the enormous value of opium in medicine, the northern Europeans seem never to have grown it as a crop plant. Yet, despite the popular notion of opium as an Oriental drug, it was from Europe that the Indians first had this poppy; and from India that the Chinese, who as late as 1550 had scarcely heard of the stuff, first imported opium and later, probably late in the sixteenth century, began to grow the poppy as a crop.

So the opium poppy is a weed here; we could have grown our own opium; but never did on anything like a commercial scale.

Tobacco is another case in point. As hundreds of amateurs growing their own tobacco to avoid paying the tax-inflated price of imported tobacco can testify, tobacco grows perfectly satisfactorily in England. The plant is a South American native which, by the time the Europeans came to America, had spread in cultivation, with the habit of smoking, throughout the whole American continent. Seeds reached Spain and France about 1550–60 and England not much later. Tobacco soon found a place in our farm flora, though a small one. It never became established here for two reasons: King James I considered smoking a filthy and dangerous habit, wrote a book against it and did his best to discourage its cultivation; and it was necessary, in order to establish our American colony in Virginia to leave the colonists the monopoly of a crop which, by sale to us, would support them.

There are other once-cultivated plants which are now quite rare. During the Middle Ages the Moors of Southern Spain introduced the cultivation of the **saffron crocus** into Europe, the first bulb reaching England late in the fourteenth century. Saffron served two purposes: as a bright yellow dye; and as a flavouring for food. The crocus was cultivated principally in Essex – hence the name of the town Saffron Walden. The growth of the synthetic dye industry and the fact that it is uneconomical to harvest – it is the stigmas and part of the styles which are used – has removed this plant from our farm flora.

Tobacco plant, *Nicotiana tabacum*

Saffron crocus, *Crocus sativus*

Broad bean, *Vicia faba*

Spinach, *Spinacia oleracea* 'Dominant'

A detail of a hop flower

Scientific Names of Plants mentioned in Chapter Four

ORIGIN		GENUS	FAMILY
N	Apple	*Malus*	Rosaceae
I	Barley	*Hordeum*	Gramineae
I	Beans (haricot, runner, French or kidney)	*Phaseolus*	Leguminosae
N	Beetroot	*Beta*	Chenopodiaceae
I	Broad bean	*Vicia*	Leguminosae
I	Broccoli	*Brassica*	Cruciferae
I	Brussels sprouts	*Brassica*	Cruciferae
I	Cabbage	*Brassica*	Cruciferae
N,I	Carrot	*Daucus*	Umbelliferae
I	Cauliflower	*Brassica*	Cruciferae
N,I	Celery	*Apium*	Umbelliferae
I	Cherry	*Prunus*	Rosaceae
N,I	Cob-nut	*Corylus*	Betulaceae
I	Filbert	*Corylus*	Betulaceae
N,I	Flax	*Linum*	Linaceae
I	Haricot (runner and French)	*Phaseolus*	Leguminosae
I	Hemp	*Cannabis*	Cannabaceae
N	Hop	*Humulus*	Urticaceae
I	Jerusalem artichoke	*Helianthus*	Compositae
N	Leek	*Allium*	Liliaceae
I	Lettuce	*Lactuca*	Compositae
I	Maize	*Zea*	Gramineae
I	Oats	*Avena*	Gramineae
I	Oil-seed poppy	*Papaver*	Papaveraceae
I	Onion	*Allium*	Liliaceae
I	Opium poppy	*Papaver*	Papaveraceae
I	Pea	*Pisum*	Leguminosae
I	Peach	*Prunus*	Rosaceae
N	Pear	*Pyrus*	Rosaceae
I	Pineapple	*Ananas*	Bromeliaceae
N	Plum	*Prunus*	Rosaceae
I	Potato	*Solanum*	Solanaceae
N	Raspberry	*Rubus*	Rosaceae
I	Saffron crocus	*Crocus*	Iridaceae
I	Spinach	*Spinacia*	Chenopodiaceae
N,I	Strawberry	*Fragaria*	Rosaceae
N	Sugar-beet	*Beta*	Chenopodiaceae
I	Swede	*Brassica*	Cruciferae
I	Sweet corn	*Zea*	Gramineae
I	Tobacco	*Nicotiana*	Solanaceae
I	Tomato	*Lycopersicon*	Solanaceae
I	Turnip	*Brassica*	Cruciferae
I	Vine	*Vitis*	Vitaceae
I	Wheat	*Triticum*	Gramineae

Chapter Five

Natives and Exotics

Woodland violets, *Viola reichenbachiana*

THE flowering plants we grow in our gardens have been 'made' by man. Wild plants are brought into cultivation, fed, tended, and only the finest selected for breeding. Different species within a genus are hybridized to realize the full potential in size, colours, hardiness, disease resistance, of that genus. If you compare the wild species with the garden varieties, it is sometimes difficult to believe that the garden plant can have come from such a beginning. In other cases the affiliation is more obvious: a garden heather, for example, is not much different from a wild one.

In this chapter the aim is to get some idea of how much of our familiar garden flora is *native* in origin, and how much is not. The best way to do it is to walk round an imaginary garden with occasional excursions into the wild country, and talk about what we see there. Having dealt with the kitchen garden in the previous chapter, we will now stick to the flower garden. There is one very big advantage in making use of an imaginary garden instead of a real one; we can have flowers of all seasons blooming at the same time. And since we need some kind of order, let it be seasonal.

So we begin with **snowdrops**; one species is native, all the others – and we have most of them established here – are introductions from southern Europe and Asia Minor. The **winter aconite**, which flowers with them, is an alien from South Europe or China; and the **dog's tooth violets** – which are not violets – are also immigrants, though real **violets** are native. We have two native species of **daffodils**, and they have contributed to the making of the big garden daffodils; but there is more foreign than native blood in them.

Nothing could be more 'natural' than the pale yellow **primroses**; and we have a native pink one, though it is rather a dingy colour; the blue ones are foreigners. As for the **polyanthus**, our native **cowslip** is its chief parent.

Bluebells are as native as primroses; but the garden bluebells, bigger if no prettier, are from Spain. As for the **hyacinths**, they come from the country round the eastern Mediterranean and were not introduced until the seventeenth century; nor were **tulips**, which came from Turkey by way of Austria. The Turkish gardeners had already transformed them by selection and hybridization before they reached us.

Most people's lawns have **daisies** on them; the little lawn daisy is a native which has given us a quite good garden daisy, white and pink and usually double. Like other garden flowers of native origin – **forget-**

Cowslips, *Primula veris*

75

me-not, some **bachelor's buttons**, some **globe flowers** – it is a more modest contribution to the horticultural picture than its foreign relations.

Among the loveliest of spring flowers are the **anemones**. We have a native one, the little **windflower**, usually white but occasionally pale blue. The more spectacular deep blues and reds are all from South Europe, introduced in the sixteenth and seventeenth centuries. The **Pasque flower**, so called because it is usually in flower at Easter, is native.

Late spring and early summer is flower time, when everything seems to bloom at once. We shall take, to begin with, three herbaceous perennials which everybody grows: paeonies, lupins, and delphiniums. There is one very small part of Britain where the common **paeony** is found growing wild – the island of Steep Holm in the Bristol Channel. This is the site of a twelfth-century monastery, and it is most likely that the monks grew this paeony there for its medicinal properties. Most of the garden varieties are descended from plants developed over centuries in China and Japan, though there are some modern American hybrids also.

Delphiniums have originated largely from *Delphinium elatum*, a wild plant of central and southern Europe and Asia as far as the Caucasus, which has been cultivated in England since 1578 and was hybridized with other species about a century ago. Delphiniums are perennial and thus differ from the annual larkspurs, now usually called *Consolida* by botanists, which come from southern Europe and at one time were naturalized in some English cornfields.

The garden **lupins** are derived from a North American species introduced in 1826. There are, in the wild, blue, white and pink varieties; hybridization has produced the extraordinary range of colours we now enjoy in this flower.

Marigolds: a member of the genus *Calendula*, introduced from the Mediterranean region in the sixteenth and seventeenth centuries.

Columbines, as we grow them in gardens, are very much 'made' flowers. We have a native species and it has contributed to the making of the short-spurred hybrids; the long-spurred hybrids derive from a Canadian species, and other contributors to our garden hybrids have come from the Rocky Mountains (1864), New Mexico (1873), West Canada and elsewhere.

The garden **snapdragons** were introduced from South Europe – the date of introduction does not seem to be on record – and here, again, we have a case of a garden introduction enriching the wild flora, for snapdragons can be found well established and naturalized in many parts of Britain.

How about all the garden **pinks** and **carnations**? These are 'made' plants; in the rock-garden group the plant makers have used one English native species and other species from the Alps and Central Europe. The true carnations are from species from the south of France; but they have been crossed with pinks to produce a whole range of garden plants, the group named after the nurseryman Montagu Allwood. The 'perpetual' greenhouse carnations are half-French and half-Japanese in their genetic composition.

The **foxglove** is an extremely common native plant, particularly

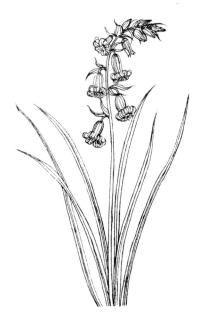

Bluebell, *Hyacinthoides non-scripta*

Water forget-me-not, *Myosotis scorpoides*

abundant in the West Country but very widely distributed in our flora.

The big, border **irises** are the creation of British, French, German and American plant breeders out of scores of varieties of dozens of species from all over Europe, Africa, and Asia. There is nothing either natural or native about them. The same can be said, but with reservations about the garden **lilies**; reservations because some of those we grow are the unchanged species – the tiger lily, the regal lily and the golden ray lily are, for example, doubtless larger in gardens than in the wild, but they are the same plants. Have we any native lilies? It is very doubtful. Two species of the turk's cap group can be found growing unquestionably wild in several parts of Britain. But they are almost certainly naturalized rather than natural here. Though well-established and persistent, they are confined to very small areas and seem unable to spread beyond them.

The **mulleins** are splendid ornaments of the summer garden. We have a native one – the hag-taper or hedge-taper – but the garden mulleins are from species introduced from South Europe and the Middle East.

There are very numerous daisy-shaped flowers in most gardens, white, red, yellow, blue and even in shades of brown. The big, robust white daisies come from either Portugal or the Pyrenees. The true **marguerites** which used to be bedded out with 'geraniums', come from the Canary Islands. Although we commonly reserve the name chrysanthemum for the autumn-flowering kinds, all these and scores of other daisies belong to the genus *Chrysanthemum* and have reached us from every corner of the world. The red, orange and mahogany **gaillardias** come from North America, the **gazanias** from South and Central Europe, though they are naturalized here. The common **sunflower** is North American, introduced in the seventeenth century or perhaps, late in the sixteenth. The **African daisies** (*Arctotis, Gerbera*, etc.) are from South Africa. The so-called **French** and **African marigolds** are both seventeenth or eighteenth century introductions from Mexico.

We can move on to the late summer daisy flowers while we are on the subject and return to look at some others later. The **China asters** come from China and Japan. The vast range of **michaelmas daisies** are mostly derived from North American species and were not seen in England before the sixteenth century; they are now so hybridized as to be a completely artificial group of garden plants. Our border and our greenhouse **chrysanthemums** did not reach us until the eighteenth century and then not as species but as complete hybrids, the product of centuries of work by Chinese and Japanese gardeners on species native in their respective countries.

The flowers we call '**geraniums**', but which are in fact *Pelargonium* species, come from South Africa and other parts of the southern hemisphere and we have only had them since late in the eighteenth century and most of them since the early nineteenth century. The **arum lilies** arrived at the same time, from South Africa. **Aubrietas** are from Greece. All the **begonias** come from the tropics and subtropics, and are of relatively recent introduction. A couple of **bell-flowers** are native, but all the best garden kinds are bred from European and Asian species introduced during the past three centuries, and *most* of them relatively

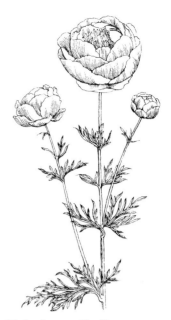

Globe flower, *Trollius europaeus*

Pasque flower, *Pulsatilla vulgaris*

Above left: snowdrops are woodland flowers; *above right:* the Pasque flower grows on chalk downs; *below:* chalkland beech wood with bluebells, Dorset. *At right:* a spring garden with anemones, daffodils, daisies, dog's tooth violets, hyacinths, primulas, pasque flowers, snowdrops, and winter aconites.

recently. The **calceolarias** are South American, introduced as species in the nineteenth century and hybridized by English plant-breeders.

Three genera of herbaceous plants tend to dominate our flower gardens in late summer and autumn. Chrysanthemums and Michaelmas daisies we have already noticed; **dahlias** have become more important than either of them.

There are about a dozen species of dahlias and they are found wild in only one country in the world, Mexico. They were introduced to Spain in 1789: it may be that these plants, or, rather, tubers, were botanical species; on the other hand they may have been garden forms, for the Aztecs, whose empire Cortés destroyed, were great gardeners and passionately fond of brilliantly coloured flowers. Some tubers were sent to Kew in 1789, but they died and it was not till 1802 that dahlias got a footing here. There were three forms – a double purple, a single pink and a single red. In 1815 some plants of a species named, much later, after the Mexican revolutionary liberator, Juarez, reached England. So variable are all the species that the thousands – literally – of garden varieties and forms have all been developed from those few kinds. However, the point here is that we have only had dahlias in Britain for 174 years.

Another late-flowering plant we owe to Mexico is the **cosmos**. The first species reached us in 1799 and others early in the next century.

The flower we call '**nasturtium**' is not a *Nasturtium*, which is the name of the genus to which our native watercress belongs. The species was introduced from Peru in 1686. In mild winters its seeds survive in the ground and germinate so freely in the following year that the nasturtium can become a weed. I should not be astonished to hear that it was naturalized in the warmest parts of Britain.

We have found very few natives in our garden; and we are not likely to find many more however thoroughly we search. I have just been asking a small group of people to name, at random, garden flowers they like; and then identify the place they came from: **evening primrose** – North America, but quite naturalized by now; the big garden **poppies** – their homeland is Armenia. The common red poppy of the cornfields is native, of course, but the opium poppy which can often be found growing wild is a naturalized immigrant from south-east Europe. **Petunia**? South American. The scarlet-flowering **sage** beloved of municipal gardens? Mexican. **Phlox** are from North America, introduced in the seventeenth century, and **Virginia stock** is misnamed, for it is of Mediterranean origin. **Pyrethrum** daisies come from the Caucasus mountains or from Persia; they have a double importance in the garden flora because the flower-heads, dried and ground into a powder, make a very effective insecticide which is not poisonous to human beings.

Scabious was another flower mentioned in response to my question and yet another was **penstemon**. We do have, of course, a native scabious, very common in all the limestone regions; but our garden scabious derives from South European and Caucasian species. Penstemons all come from North and Central America except for one Asian species. **Tobacco** plants – the ornamental ones – are Brazilian and **zinnias** are Mexican.

*   *   *

As we saw in Chapter Two, the English landscape has been transformed and greatly enriched by the introduction of alien trees. The introduction of alien herbaceous plants, on the other hand, has not had any noticeable effect on the countryside, despite the naturalization of numerous species. But in England the great majority of people live in cities, towns and suburbs and for them the familiar flora is the flora of parks and gardens. Remove from that flora everything which is not of native origin and the result would be an astonishing impoverishment.

Scientific Names of Plants mentioned in Chapter Five

ORIGIN		GENUS	FAMILY
I, Nt	Aconite, winter	*Eranthis*	Ranunculaceae
I	African daisy	*Arctotis, Gerbera* etc.	Compositae
N, I	Anemone (wind flower)	*Anemone*	Ranunculaceae
I	Arum lily	*Zantedeschia*	Araceae
I	Aster (China)	*Callistephus*	Compositae
I	Aubrieta	*Aubrieta*	Cruciferae
N	Bachelor's buttons	*Ranunculus*	Ranunculaceae
I	Begonia	*Begonia*	Begoniaceae
N, I	Bellflower	*Campanula*	Campanulaceae
N	Bluebell	*Hyacinthoides*	Liliaceae
I	Calceolaria	*Calceolaria*	Scrophulariaceae
I	Carnation	*Dianthus*	Caryophyllaceae
I	Chrysanthemum	*Chrysanthemum*	Compositae
I	Chinese aster	*Callistephus*	Compositae
N	Columbine	*Aquilegia*	Ranunculaceae
I	Cosmos	*Cosmos*	Compositae
N	Cowslip	*Primula*	Primulaceae
N, I	Daffodil	*Narcissus*	Amaryllidaceae
I	Dahlia	*Dahlia*	Compositae
N	Daisy	*Bellis*	Compositae
I	Delphinium	*Delphinium*	Ranunculaceae
I	Dog's tooth violet	*Erythronium*	Liliaceae
Nt	Evening primrose	*Oenothera*	Onagraceae
N	Forget-me-not	*Myosotis*	Boraginaceae
N, I	Foxglove	*Digitalis*	Scrophulariaceae
I	Gaillardia	*Gaillardia*	Compositae
I	Gazania	*Gazania*	Compositae
I	Geranium	*Pelargonium*	Geraniaceae
N	Globe flower	*Trollius*	Ranunculaceae

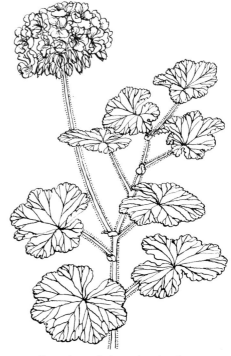

Geranium, *Pelargonium inquinans*

Left: cottage garden at Ashby St Ledgers, Northamptonshire; *above:* gaillardia, *Gaillardia aristata* 'Mandarin'; *below left:* columbine; *below middle:* lily 'Enchantment'; *below right:* penstemon. *Opposite:* the summer garden, including mulleins, dahlias, foxgloves, delphiniums, paeonies, arum lilies, nasturtiums, aubrieta, carnations and pinks

Garden pink, *Dianthus allwoodii*

ORIGIN		GENUS AND SPECIES	FAMILY
N	Hag-taper	*Verbascum*	Scrophulariaceae
N	Hedge-taper	*Verbascum*	Scrophulariaceae
I	Hyacinth	*Hyacinthus*	Liliaceae
N, I	Iris	*Iris*	Iridaceae
I	Larkspur	*Delphinium*	Ranunculaceae
I, Nt	Lily	*Lilium*	Liliaceae
I, Nt	Lupin	*Lupinus*	Leguminosae
N, I	Marguerite	*Chrysanthemum*	Compositae
I	Marigold	*Calendula*	Compositae
I	Marigold (French, African)	*Tagetes*	Compositae
I, some Nt	Michaelmas daisy	*Aster*	Compositae
N	Mullein (hag-taper, hedge-taper)	*Verbascum*	Scrophulariaceae
I	Nasturtium	*Tropaeolum*	Tropaeolaceae
I, Nt	Paeony	*Paeonia*	Ranunculaceae
N, I	Pasque flower	*Pulsatilla*	Ranunculaceae
I	Penstemon	*Penstemon*	Scrophulariaceae
I	Petunia	*Petunia*	Solanaceae
I	Phlox	*Phlox*	Polemoniaceae
N, I	Pink	*Dianthus*	Caryophyllaceae
I	Polyanthus	*Primula*	Primulaceae
N, I	Poppy	*Papaver*	Papaveraceae
N	Primrose	*Primula*	Primulaceae
I	Pyrethrum	*Chrysanthemum*	Compositae
I	Sage	*Salvia*	Labiatae
N, I	Scabious	*Scabiosa*	Dipsacaceae
I	Snapdragon	*Antirrhinum*	Scrophulariaceae
N	Snowdrop	*Galanthus*	Amaryllidaceae
I	Sunflower	*Helianthus*	Compositae
I	Tobacco	*Nicotiana*	Solanaceae
I, Nt	Tulip	*Tulipa*	Liliaceae
N	Violet	*Viola*	Violaceae
I	Virginia stock	*Malcolmia*	Cruciferae
N, I	Windflower (anemone)	*Anemone*	Ranunculaceae
I, Nt	Winter aconite	*Eranthis*	Ranunculaceae
I	Zinnia	*Zinnia*	Compositae

Chapter Six
Transformation by Hybridization, Mutation and Selection

HYBRIDIZATION, mutation and selection are the names given to three processes by which plants can be transformed to create new kinds not found in nature.

Hybridization

The physical attributes of plants are governed by bodies called genes. If both the male (pollen) parent and female (seed) parent of a plant belong to the same species, it will – except in the rare case of a mutation – be identical with its parents in all its characters: size, colour of flower, shape of leaves, habit of growth, etc. But if the pollen from one species is used to fertilize the flower of a different species, then a mingling of genes occurs and the offspring of this *hybridization* (or in other words, cross-breeding) will have characters intermediate between those of the two parents. As genetic characters can be combined by hybridization in an enormous number of different ways, the seedlings resulting from hybridization will not all be the same. So the plant-breeder has a choice of seedlings with different flower colours, vigour, scent, etc., from a single cross-breeding operation.

As a rule only species *within a single genus* are capable of cross-breeding. You can, for instance, cross the dog rose with the burnet rose; but not a rose with an apple – even though they belong to the same family. There are, however, a few genera so closely related that inter-generic hybridization is possible. Hybridization between genera in different families is never possible.

Mutation

From time to time, very rarely, a change occurs in the gene construction of an individual plant. As the genes govern the characters of plants that individual plant will look different from the others of its species. For example, if the species normally has red flowers, the plant which has undergone genetic *mutation* may have white flowers or yellow flowers. Such mutations occur naturally and, if the new characters are such as to favour survival, give rise to new species. Mutations can also be provoked, for example, by treating seeds with a poisonous substance called colchicine, derived from the autumn crocus (*Colchicum*); or by irradiating seeds with X-rays.

Selection

A plant-breeder making new kinds of plants by hybridization (or, much more rarely, mutation) will have scores, hundreds or even thousands of seedlings, resulting from each single operation, in his trial beds. For, although he can be sure of getting some of the new characters he wants, or nearly so, by careful choice of parent species, no two seedlings will be the same. He now, therefore, selects those which have the best characters – colour, scent, vigour, flavour, disease-resistance, etc. – scraps all the rest, and either goes on to more hybridization among the selected seedlings, or tries to fix the good characters of seedlings by self-fertilization.

* * *

Taking a small number of genera (roses, daffodils, lilies and tulips) by way of example, it should now be possible to give some idea of how the floral aspect of England has been transformed not simply by the introduction and mass establishment of alien plants, but by the processes described above.

Roses

There are at least a hundred species of wild roses, growing all over the northern hemisphere, and some of which we grow in gardens as they are; but most garden roses of today are the product of an amazing amount of breeding and selection within the past two centuries. Before that, popular varieties were usually found accidentally, as seedlings or natural hybrids of existing species and varieties, and this process goes back well over 2000 years. It is a remarkable fact that probably only four or five true species are involved in almost all the hybrids, which started to appear in increasing numbers from the sixteenth century.

A new impetus to rose breeding began in 1792 when a species of rose introduced from China – till then cut off from the rest of the rose world – was used for breeding, to be followed shortly by three others, all hybrids from the same species, the China rose. Crossed with European hybrid roses, these gave rise to a new array of garden varieties, among which a few bloomed for more than one brief period each year. Eventually roses were produced which bloomed much of the summer and were rather grandly called hybrid perpetuals.

Roses of the last century were often flattish, multi-petalled and strongly scented, and the painting opposite shows a group of them. Although the first 'modern' rose – the beautifully formed hybrid tea we see in every garden today – was actually raised in 1867, it was a long time before the 'old-fashioned' type, as we call their survivors today, were largely ousted from gardens. This was partly due to a change in taste, and partly because the older roses do not always have strong constitutions.

Although hybrid teas and the usual less formal, cluster-headed floribundas are the favourites today, breeders are perhaps getting a little tired of them and some are certainly aiming for blooms of different

Roses: 1 Common moss (*Rosa centifolia muscosa*). 2 La Reine Victoria (Bourbon). 3 Cuisse de Nymphe (alba). 4 Buff Beauty (musk). 5 Albéric Barbier (wichuraiana). 6 Conrad F. Meyer (rugosa). 7 Charles de Mills (gallica). 8 Hebe's Lip (damask). 9 Austrian copper briar (*Rosa foetida*). 10 Fantin-Latour (centifolia). 11 Félicité et Perpétue (sempervirens)

forms in the rose, probably the most popular and widely grown garden flower of all time.

Daffodils

Daffodils belong to the genus *Narcissus* of which there are about seventy species. At least two are natural to our scene – the wild daffodil *Narcissus pseudonarcissus*, and the Tenby daffodil, *N. obvallaris*. But in various parts of the world there are other species which belong to the same group within the genus: most of these come from the Pyrenees, though there are some from Portugal and northern France, and a few of unknown provenance. With this raw material were created the enormous daffodils which, to most of us, the majority who have never seen a wild daffodil, *are* daffodils.

The work of deliberate hybridization which transformed the image of what we mean by 'daffodil' did not begin until 1820 when a clergyman called Herbert began crossing the various species. Those who followed his example used *N. hispanicus* (introduced to England from northern France and Spain before 1600) as the chief source of pure yellow hybrids and *N. bicolor* to get daffodils with a yellow perianth and white trumpet. But why are the cultivated daffodils so much larger than the wild ones?

Top row: hybrid narcissi.
Left to right: trumpet daffodil ('King Alfred'); tazetta narcissus ('Grand Soleil d'Or'); large-cupped narcissus ('Brahms'); cyclamineus narcissus ('Charity May'); double narcissus ('Tahiti')
Bottom row: wild narcissi. *Left to right:* hoop petticoat daffodil, *N. bulbocodium*; paper-white narcissus, *N. tazetta papyraceus*; wild daffodil, *N. pseudonarcissus*; jonquil, *N. jonquilla*; pheasant's-eye narcissus, *N. poeticus*; *N. tazetta*

Some of the answers to this question are so complex and require so considerable a knowledge of genetics for their understanding that all I can do is to give a general idea of them; others are simple enough. First and simplest, there is the process, already described, of selection. Then, cultivation alone tends to increase the size of plants, flowers and fruit; in nature a wild plant has to compete with others of its own and other genera for nourishment and water. In the garden or plantation it is protected from competition, watered and fed, and so can realize its full growth potential.

The phenomenon called gigantism is another cause of increase in size; it is common in food plants but uncommon in ornamentals. The nature of the phenomenon is clear enough: every cell in the tissue of the giant plant is two or three times larger than the cells in a normal plant of the same species. But there has never been a satisfactory explanation of why this happens.

Finally and most important there is the genetical phenomenon called polyploidy. In every cell of a plant (or any living creature) there is a nucleus which is, as it were, the cell's working part; in that nucleus are a number of rod-like bodies called chromosomes and in all normal plants there is always an even number of these. Growth is caused by cell division; when a cell is going to divide there occurs a process called mitosis. Each chromosome splits longitudinally into two, so that if there were ten chromosomes per cell, there are now twenty, but only half the size of the original ones. At this stage the split chromosomes are called chromatids; the chromatids now move outwards away from each other towards the 'walls' of the cell; a new cell 'wall' grows between them; the chromatids grow into full-size chromosomes; where there was one cell, there are now two, each with ten chromosomes as before.

Now it sometimes happens that there is a temporary breakdown in this process and after the chromosomes have split into chromatids no new cell wall forms to divide them. But the chromatids still grow into chromosomes. So now we have cells which instead of having ten chromosomes have twenty. And the cells are, of course, bigger and so, therefore, is the plant or that part of it in which this breakdown has occurred. The normal process then being resumed, instead of ten chromosomes splitting into twenty chromatids to form two cells with ten chromosomes each, we have nuclei of twenty chromosomes splitting into forty chromatids which gives us two cells, each, again, with twenty chromosomes.

The basic number in a single set of chromosomes is called the haploid number. In normal plants every cell has *two* sets of chromosomes; so they are called, in genetics, *diploids* (di = 2); but if, by the process I have sketched here above, the cells of a plant or part of a plant have not 2 × the haploid number but 4 × the haploid number, that plant is called a tetraploid. The phenomenon of haploid number multiplication is called polyploidy; when polyploidy occurs size increases. And as it is very often a genetical mutation occurring in connection with hybridization, it is commoner in cultivation than in the wild.

Wild or species lilies, which include some
of the parents of present-day hybrid lilies.
Also shown is the 'giant lily' (*Cardiocrinum
giganteum*), a close relation which can grow
four metres tall.
Back row, left to right:
Lilium sulphureum and *Cardiocrinum giganteum*
Middle row, left to right:
L. *henryi, candidum, speciosum, regale* and
hansonii
Front row, left to right:
L. *dauricum, martagon album, japonicum,
tigrinum, rubellum, chalcedonicum, croceum* and
cernuum

Lilies

Lilies, of which there are around ninety species, come from Europe, many parts of Asia, Japan and North America. It is not surprising that their actual habitat varies enormously. They range from sea level to Himalayan mountains up to 12,000 feet; some grow in such unlikely sites as coral reefs and on volcanic ash and lava. Some live in almost desert conditions on the steppes of Russia and Manchuria, others flourish in hot monsoon climates.

While it is remarkable how many species have been grown in gardens, they are not in general very tolerant of conditions far different from those they are accustomed to, and most dwindle away in a few years unless given special care. Moreover, they are readily infected with a deadly fungus disease and also a group of incurable diseases called viruses, neither of which are usually found in the wild, but rapidly infect and gradually kill lilies in cultivation.

The hybridization of lilies can as it were 'iron out' demands for special growing conditions found in the species. Thus, if one crosses a lily preferring a chalky soil with one preferring the opposite, acid soil, the probability is that some of the offspring will be tolerant to both types of soil.

Another advantage of breeding is that offspring from two distinct species often have what is called 'hybrid vigour'. Though the reason for this is rather obscure, there is no doubt that the first new generation from such a cross – known as an F.1 – is often much more robust than either of its parents, and many modern annual flowers and vegetable seeds in particular are produced by this method. With lilies, annual cross-breeding, and growing on the seedlings into flowering-size bulbs over two or three years, has the special advantage that virus diseases are not transmitted by seed, so that the bulbs are guaranteed healthy.

By such methods we now have strains of hybrid lilies with all the flower shapes of the various species, often with flowers of great size, and in a remarkable range of colours.

Tulips

There are, in nature, about 100 species of the genus *Tulipa*, native to Europe, West Asia, Central Asia and North Africa. But it was not in our gardens or in any European gardens that this genus was originally transformed and the vast range of enormous garden tulips produced. That work was begun in Turkey, and when the first garden tulips reached England flower-lovers received them as something new which they had never seen before. Thus, for example, in 1582 Richard Hakluyt was writing: 'Within these four years there have been brought to England from Vienna in Austria divers kinds of flowers called Tulipes . . . procured a little before from Constantinople by an excellent man called M. Carolus Clusius.'

So great has been the transformation of this genus that no known species can be regarded as the ancestors of the garden tulips. Richard Gorer in his excellent work *The Development of Garden Flowers** suggests that the original ancestor was collected to such an extent that it became extinct in the wild.

Whether the story begins in Persia or Turkey is not clear. Two of the greatest Persian poets, Hafiz and Omar Khayyam, mention tulips in their poetry but it is not possible to tell whether they refer to wild or cultivated kinds. But it was certainly from Turkey that we first had the bulbs: by the mid sixteenth century there were over 1300 varieties in cultivation in the gardens of the Ottoman Empire. Between 1702 and 1720, in the reign of the Sultan Ahmed III, there was a tulip mania with bulbs of new, especially fine varieties fetching such extortionate prices (just as had occurred in Holland between 1623–7) that the Sultan had to take a hand and fix maximum prices by decree.

In 1554, Ghiselm de Busbecq, the Imperial Ambassador at the Sublime Porte when Suleiman the Magnificent was Sultan was travelling through Adrianople to Constantinople, and there saw tulips in gardens. He remembered them, procured bulbs or seeds and sent them to the Fuggers, the great bankers of Augsburg, who had the grandest gardens

*Eyre & Spottiswoode, 1970.

Top row: hybrid tulips. *Left to right:* cottage tulip ('Mrs John Scheepers'); Rembrandt tulip ('May Blossom'); Darwin tulip ('La Tulipe Noire'); parrot tulip ('Estella Rynveld'); lily-flowered tulip ('Inimitable'). *Bottom row:* wild tulips. *Left to right:* T. maximowiczii; T. fosteriana; T. whittallii; T. clusiana; T. kaufmanniana; T. kolpakowskiana*

in Europe – or, at least, florally the grandest – and it was in those gardens that Conrad Gesner saw the tulips in 1559. De Busbecq also sent tulip seeds to Clusius, who was working in Vienna, and Clusius sent some to England. When, in 1593, he became Professor of Botany at Leyden University he planted tulips there; they were stolen and became the origin of the great Dutch tulip industry.

Now a curious thing: the Turkish ideal of a perfect tulip flower was a self-coloured one with pointed petals. Tulips are subject to a phenomenon (long a mystery but now known to be due to a virus infection) called 'breaking' in which the flower is streaked with colour on a yellow or white ground. Such 'broken' tulips were rejected by Turkish gardeners but preferred by the English until the fairly recent revival of a taste for self-coloured varieties. This taste for 'broken' tulips led to the propagation of those with especially clear markings or favoured colours, until the diversity of kinds was enormous and the garden tulips, in colour, form, markings and size, were like no tulip in nature.

A bed of 'broken' tulips in Regent's Park, London

Chapter Seven

The Heroes of Plantsmanship

WHO were the people who, by bringing or sending foreign plants to Britain, so changed its aspect that a prehistoric Briton would not recognize more than one in ten of the plants he would see in a day's walk if he could return to present-day Britain?

For the really early cases we have, of course, no names. Those great benefactors who first brought the seeds of wheat and barley and beans and peas to England must be for ever anonymous. But we are chiefly concerned with plants which were introduced because they are beautiful, not those that are useful; when, and by whom, was the task of introducing them started?

It is possible that this work began before the conquest of Britain by the Roman Emperor Claudius in the first century A.D. Roman civilization had penetrated to the Belgic tribes in the south and south-east of Britain before the conquest, and the chieftains at least were familiar with such luxuries as wine. But if introduction of new plants other than the wheat and barley of the Neolithic farmers did occur this early, we have no means of knowing it; it is safer to assume that the earliest introductions were made after the conquest.

The reasons why the Romans set about adding to our flora are obvious. Some plants they wanted in their new province for their usefulness: plants such as fruit trees and fruit bushes and medicinal herbs. But there was more to it than that: the military officers and civil servants sent on duty to Britannia felt very far from home – were, during their term of service, exiles. They had a natural wish to recreate round them, especially in their villas, the familiar Italian or Gallic scene. A Roman garden was a formal creation and its elements were fixed; to make their gardens, the owners of villas and builders of towns brought from France or Italy those familiar and beloved plants necessary to 'Romanize' the scene as far as that was possible in our climate, just as the British officers and civil servants who used to administer the Indian Empire did their best to grow familiar English flowers in their gardens.

There is, however, one very great difficulty in knowing for certain what all their introductions were, because of the period of confusion and war which followed the Roman withdrawal. Plants which were naturalized here by the fourth century survived; those which still needed care did not. We know, for example, that the Italian cypress, an important element in the Roman landscape, was in cultivation here in the fifteenth century. Had it recently been re-introduced or had it survived from Roman times? Nobody can say for sure; and the same may be true

Two frescoes showing Roman gardens: *above*, from the villa of the Empress Livia at Prima Porto; *below*, from a house at Pompeii

95

Holm oak, *Quercus ilex*

for certain poplars not native here, and of the evergreen oaks. Because one must rely on sure records and not on guesswork, I have, in Chapter Three, assumed that a genus of trees was not to be found in England until the first recorded date; but I am inclined to think that it was the Romans who added cypress and holm oak to the number of our native trees.

One woody plant we can be sure about is the grape-vine (*Vitis vinifera*). Although villa owners must have wanted to try it, it was not introduced for more than a century after the conquest, for a particular reason. The Italian wine industry bosses had become so alarmed at the inroads which wine-growers in the provinces were making into their profits, and the Roman government at the extension of vineyards at the expense of grain-fields, that a law forbidding the planting of vines in the provinces and requiring provincials to dig up their vineyards was passed through the Senate. It was not repealed until A.D. 280 (by the Emperor Probus). When it was, there was a transformation in the English scene: vineyards were planted, chiefly in Kent, Norfolk and Gloucestershire. And they remained familiar in the cultivated landscape for well over a thousand years. How do we know the vines survived the Roman withdrawal and were not re-introduced? Because the first reliable post-Roman English historian, the Venerable Bede, writing in the eighth century, mentions the vine as a familiar English plant.

Typical European vineyards of the late fifteenth century, from a manuscript in the British Library

Another newcomer we owe to the Romans is the fig tree, *Ficus carica*: it was introduced, of course, for its fruit but it added a new note to the visual aspect of foliage in Britain. Planted first in the south it spread north until it was established in Yorkshire; it has flourished here ever since.

Apple and pear were probably cultivated here before the Roman occupation. But the sweet cherry is one of their introductions. There are no records of the introduction of such Mediterranean shrubs as rosemary, lavender, lavender cotton, culinary thyme and sage, and in Chapter Four I have taken the safe course of assuming them to have been introduced into our cultivated flora by the monastic gardeners. But it is quite possible that they have all been here since the second or third century.

There are scores of species of plants which we regard as native because they are established in our wild flora, but which may not have been an element of the flora with which man was presented in England. A good example is the weed, more familiar by the seaside than inland, *Valeriana officinalis*. The medicinal qualities of this plant were much valued in Roman Italy. It is more than possible that valerian, planted here by Roman physicians, only became a part of our flora some time during the Roman occupation.

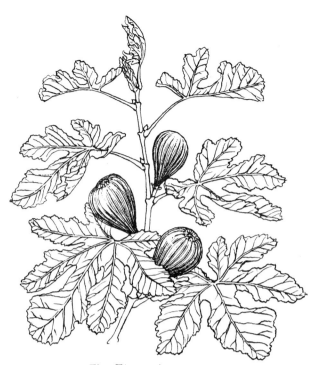

Fig, *Ficus carica*

*　　*　　*

While it is inconceivable that there were no changes wrought by plant introductions during the ten centuries following the Roman withdrawal, it is impossible to do more than guess at what those changes were. There are some culinary plant introductions we can be sure of: carrots reached us in the fourteenth century – but as we have noted they were purple and we did not have the familiar orange-coloured carrot until the eighteenth century; cabbage we had had before the Romans came, but spinach not until the sixteenth century. Onions, the Romans brought us, and perhaps the turnip, and certainly they grew asparagus here though we may have lost it when the occupation ended; celery, on the other hand, was a seventeenth century or even early eighteenth century newcomer. However, we are not in this chapter concerned with the kitchen-garden flora. The first great age of plant introduction of the kind which changes the visual aspect of a country was that of the Tudors and Stuarts. And the real pioneers were the Tradescants, father and son.

John Tradescant the elder was a Suffolk man who became head gardener to the first Lord Salisbury – Queen Elizabeth the First's Secretary Cecil. In 1611 he began plant-hunting travels which took him to Holland and Flanders, parts of Russia and Algeria. Later, after his return, he became head gardener to Charles the First and was succeeded in that office by his son, also named John, whose plant-hunting took him to North America. The Tradescants grew the new plants which they brought back from foreign parts in their Lambeth garden.

Of the many plants which were unseen in England until the Tradescants introduced them, I select three genera as having so enriched our garden flora as to alter its aspect: lupins, Michaelmas daisies and Virginia creeper.

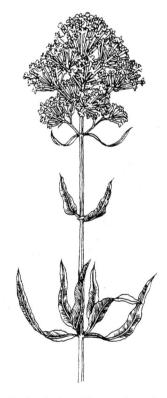

Red valerian, *Centranthus ruber*

Above: John Tradescant's house in South Lambeth which contained his famous plant collection, as it was in 1798.
Below left: John Tradescant the elder. *Below right:* John Tradescant, the younger

The same period produced a number of plant-hunters whose work, though valuable, does not really concern us here because it was confined to Britain. They studied the wild English flora and brought some of our native plants into cultivation, but they did not enrich it with new plants from abroad. Such were the botanist John Parkinson and the herbalist and apothecary Thomas Johnson, both at work early in the seventeenth century. Their job was to make us aware of what we had, not to add to it. William Turner was the (Tudor) forerunner of this kind of plant-hunter and John Ray, who died in 1705, the greatest.

Among the other sort, the introducers of alien plants, Henry Compton, Bishop of London under Charles II, was outstanding in the seventeenth century. He did not himself go off to foreign parts to look for new plants but as our North American colonies were a part of his diocesan responsibilities he was able to set his missionary and parish clergy in the Americas to plant-hunting for him and then to grow what they sent to him in his Fulham Palace gardens.

Far left: Henry Compton. *Left:* John Ray
Above: statue of John Parkinson
Below left: the Bishop of London's palace at Fulham, 1788. *Below right:* the famous cork tree in the gardens of the palace, planted in 1680 by Henry Compton when Bishop of London

'The Kitchen Garden'
page 497 of John
Parkinson's *Paradisi in
Sole*, 1629. The caption
reads:
'1 *Maluacrispa* French
Mallowes. 2. *Endivia*
Endive. 3 *Cichorium*
Succory. 4 *Spinachia*
Spinach. 5 *Lactuca crispa*
Curld Lettice. 6 *Lactuca
patula* An Open Lettice.
7 *Lactuca capitata vulgaris*
Ordinary cabbage
Lettice. 8 *Lactuca capitata
Romana* The Great
Roman cabbage Lettice.
9 *Lactuca Aquina* Corne
Sallet or Lambes Lettice.'

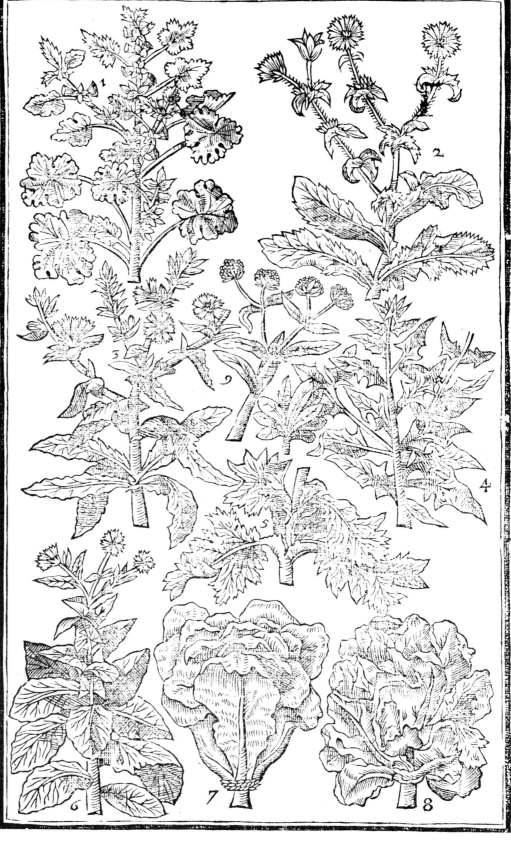

Early in the eighteenth century two men, one in America and the other in England, were together responsible for introductions which again changed our floral aspect. John Bartram was a Quaker, son of an emigrant who had settled in Pennsylvania in 1680 and who, revolting against the Friends' Society's social discipline, migrated again, to North Carolina where he was killed in a fight with Yamasee Indians. John, his son, was an exceptionally successful farmer on a piece of land which he bought in 1728 on the banks of the Schuylkill River; he had been interested in plants from the age of ten and he was the first American to lay out and plant a botanical garden.

Above: Peter Collinson. *Left:* John Bartram's house on the Schuylkill River in Pennsylvania, U.S.A.

The Englishman in this partnership was a London merchant trading with the colonies, Peter Collinson, an amateur botanist and keen gardener who was always nagging his American correspondents to send him plants. One of them must have put him in touch with Bartram. They came, after a tentative start, to an arrangement whereby Bartram sent Collinson boxes of plants and seeds, Collinson financing Bartram as a full-time plant-hunter by creating a syndicate of botanists and gardeners who subscribed £10 a year each and received a share of whatever Bartram sent to England. The arrangement worked quite well for over thirty years. Bartram was appointed by means of Collinson's syndicate Botanist Royal with a salary of £50 a year. He penetrated deep into the unexplored parts of North America and seems to have been able to get to terms with the Red Indians although he openly despised them and their ways.

Between them, Bartram and Collinson introduced about two hundred species of American plants into the cultivated flora of England; trees, shrubs and herbaceous plants. Perhaps the most spectacular was *Magnolia grandiflora*.

Magnolia grandiflora

101

Two adventurous seafaring botanists and naturalists were at work in the southern hemisphere in the mid eighteenth century: the Frenchman Philibert de Commerson, naturalist aboard de Bougainville's *La Boudeuse* during her voyage of exploration; and Joseph Banks, naturalist aboard Captain Cook's *Endeavour*. De Commerson probably discovered and described more new plants than any other single botanist has ever done but a great many of them were tender. Banks was a rich man, the doyen of the scientists of his age – he was President of the Royal Society for forty-three years – and he was the real first maker of Kew Gardens, at that time still the private property of the Royal Family. Banks not only went plant-hunting himself, but trained and sent out young botanists to bring home new plants from the West Indies, Ceylon, St Vincent and other remote places. He himself collected plants in Southern Patagonia, in the Malay Archipelago, in Newfoundland, and Australasia. One of his 'monuments' is the superb climbing rose, *Rosa banksiae*, named in his honour. He was responsible for Captain Bligh's expedition to transplant breadfruit from the Pacific islands to the West Indies as cheap food for slaves, which ended in the famous mutiny and Bligh's almost incredible voyage with a few loyal members of the ship's company in an open boat.

Above: Philibert de Commerson. *Below:* Sir Joseph Banks. *Right: Rosa banksiae*

Above: gathering specimens of the bread-fruit tree; the object of Captain Bligh's famous expedition. *Left: Artocarpus communis,* the uru or breadfruit, found in Tahiti and painted by Sydney Parkinson in 1769

Above: Francis Masson. *Below:* David Nelson, the *Bounty*'s gardener, was one of those set adrift with Captain Bligh after the mutiny

Among Banks's pupils the most successful was Francis Masson, a Scot trained at Kew; he collected plants in Spain, Portugal, South Africa, the Canary Islands and Madeira, the Azores and the West Indies. His last expedition was to North America in 1805 where, in one of the hardest winters of the century, he died of exposure. One of the changes wrought by his work in the garden aspect of England was the introduction of the genus *Pelargonium* – commonly called 'geraniums'. His adventures included being hunted by an escaped chain-gang of convicts in South Africa – they wanted him as a hostage; and being twice captured by the French, once in New Grande and once, at sea, by a privateer.

Another of Banks's young men, David Nelson, died of pneumonia in the course of Captain Bligh's open-boat voyage of 4000 miles following the *Bounty* mutiny.

The first European botanist to spend time in Japan and study its plants, though under strict constraints, was a German, Engelbert Kaempfer. That was in the mid seventeenth century. The Swedish botanist C. P. Thunberg did some useful collecting there in the late eighteenth century. The Macartney mission to the Chinese emperor in 1791 included two gardeners and the plants they collected were the mission's only useful result. British ships' masters, trading to Canton and Macao were able to buy Chinese garden plants in nurseries and take them to England – introducing camellias, azaleas, paeonies and chrysanthemums, and Banks sent a gardener named Kerr to Canton. The East India Company's Inspector of Tea at Canton, John Reeves, em-

CAROLI PETRI THVNBERG
MED. DOCT. PROF. REG. ET EXTRAORD. ACADEM.
CAES. N. C. REG. SCIENT. HOLMENS. SOCIET. LITTER.
VPSAL. PATRIOT. HOLMENS. BEROLIN. N. SCRVT.
LVNDIN. HARLEMENS. AMSTELDAM. NIDRO-
SIENS. MEMBRI

FLORA
IAPONICA
SISTENS
PLANTAS
INSVLARVM IAPONICARVM
SECVNDVM
SYSTEMA SEXVALE EMENDATVM
REDACTAS
AD
XX CLASSES, ORDINES, GENERA
ET SPECIES
CVM
DIFFERENTIIS SPECIFICIS, SYNONYMIS PAVCIS,
DESCRIPTIONIBVS CONCINNIS ET
XXXIX ICONIBVS ADIECTIS.

LIPSIAE
IN BIBLIOPOLIO I. G. MVLLERIANO
1 7 8 4.

C. P. Thunberg, *above*, and, *right*, the title page of his book *Flora Japonica*, published in 1784

ployed Chinese to hunt for plants and seeds in the interior and send them home to Kew, and another collector responsible for sending valuable Chinese plants home to Europe was the botanist Alexander von Bunge, one of the scientists attached to the Imperial Russian Ecclesiastical Mission in Peking.

As for Japan, another German, Philipp von Siebold, was the first to send any considerable number of its plants to Europe. When von Siebold first went to Japan it was almost impossible to get into the country at all, let alone travel it freely. The ruling military dictators, Shoguns, did not want the people they ruled and oppressed to have any contact with foreigners, least of all with Western 'foreign devils'; it might give them ideas about justice and liberty. But von Siebold had a special advantage over other foreigners: he was a very good oculist and ophthalmic surgeon; and the Japanese suffered from bad eyesight and diseases of the eye. So he was allowed to travel about Japan and, while treating the people for their eye troubles, was able to collect plants and send them back to Leyden University in Holland, and from there they reached England. Von Siebold was not, himself, a Dutchman; he was a German aristocrat; but Leyden had the best botanical garden and

Azalea 'Hugh Wormald'

105

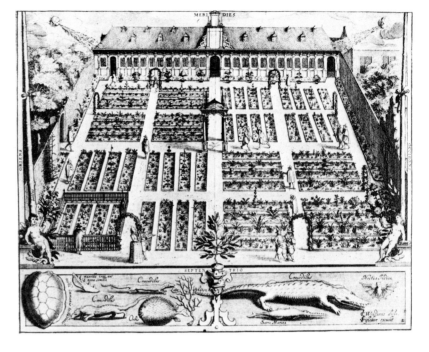

Above left: an 'affectionate' drawing of the young von Siebold from a Japanese manuscript. *Below left:* the bust of von Siebold in Leyden Botanic Gardens. *Right:* the Botanic Gardens at Leyden, *above*, in the sixteenth century and, *below*, in 1866

botanists in the world at that time. The Japanese, by the way, have not forgotten him and his services as a doctor; when parties of them visit the Leyden Botanic gardens today, they go and bow ceremoniously to the bust of their one-time benefactor, and ours, Philipp von Siebold.

The first of the great nineteenth-century collectors was David Douglas, a Scot born at Scone. Apprenticed as a garden boy in the Earl of Mansfield's gardens at Scone and later appointed gardener in the then famous Valleyfield Gardens at Dumfermline, he taught himself botany and obtained a post at the Glasgow Botanical Garden. He came to the notice of the Professor of Botany at Glasgow University, William Hooker, later Sir William, and director of Kew Gardens, and, as a result, was sent, then aged twenty-four, to collect plants for the (London) Horticultural Society in North America. That was in 1823. His first trip was to the east coast, but on later ones he went to the west, getting there by sea via Cape Horn, and he was in virtually unexplored country for most of the time. His successes were fabulous – to him we owe our finest conifers and a large number of other trees, shrubs and herbaceous plants. He died in 1829 while plant-hunting in the Sandwich Islands, by being gored to death by a wild bull when he fell into a pit-trap into which the animal had fallen before him. It has been suggested that this horrible death was no accident but a way of committing suicide. Douglas was a very religious man who had been taught that every word of the Bible was literally true. When he read Charles Darwin's *Origin of Species*

John Reeves

Below left: Douglas firs planted in the New Forest in 1860. *Below:* David Douglas

the foundations of his faith crumbled away, for he was too good a scientist not to recognize the truth when he saw it. He is supposed to have found life without religious faith unbearable. I must say I do not believe this story; poor Douglas died by a shocking accident; no man, however depressed, would choose such an atrocious and grotesque way of dying.

Above: Doctor Nathaniel Bagshaw Ward

Below: Ward's Closed Case used successfully in 1834 to ship plants to and from the Antipodes. *Right:* a later, very elaborate version used simply for decoration in the 1880s

In the 1830s it became easier to keep plants alive during long journeys overland or by sea as a result of Doctor Nathaniel Ward's invention of the Wardian case – a sort of sealed, miniature, and portable, greenhouse. The first plant-collector to make use of it on a considerable scale seems to have been Joseph Hooker (son of the Director of Kew Gardens, Sir William Hooker), who went as botanist on Ross's Antarctic expedition in H.M.S. *Erebus*, and who collected new plants in the Falkland Islands. The Lobb brothers, William and Thomas, also used the new cases, in South America and Java respectively.

We are not here concerned with the collectors of orchids and other plants for hot-houses, in the tropics, who flourished at this time. Collectors working in the tropics of the New and Old Worlds brought home hundreds of new species, but of a kind which could never be established in the open in Britain.

Joseph Hooker, on his second great adventure, collected in the Himalayas in the eighteen-forties, enormously enriching our stock of rhododendron species. This expedition was on a grand scale and Hooker had about fifty or sixty native servants – Lepchas, Bhotias and Khasias – all of whom he openly despised. He became involved through no fault of his own in Sikkimese politics and was arrested and held in prison for a month by the Rajah's *dewan* and only released when the Governor-General of British India, Dalhouse, annexed Sikkim – that is, brought it under British control. Later, after succeeding his father as Director of Kew, he made a plant-collecting expedition to the Atlas Mountains and established yet more alien species in our native scene.

The decline in the power of the Shoguns in Japan and reassertion of the imperial power which was to culminate in the restoration of constitutional power to the Mikados and the westernization of Japan, began to make Japan more accessible at the beginning of the 1860s. Von Siebold returned there in 1859. As a result a second spate of Japanese plants reached Europe, including Britain, among them many new trees, both conifers and broad-leaved deciduous trees; with them came the first bamboos hardy in our climate. Von Siebold's services to the Japanese as an ophthalmic surgeon still gave him special privileges and probably no other European could have accomplished so much at that time. For example, when the English nurseryman Veitch arrived in 1860 his movements were restricted; still, he too introduced trees new to the English scene, not to mention that grandest of all lilies, *L. auratum*, the golden ray lily.

Below right: an example of Veitch's *Lilium auratum*. *Below:* Sir Joseph Hooker's own drawing of *Magnolia campbellii*, discovered in Sikkim and named after his friend Archibald Campbell

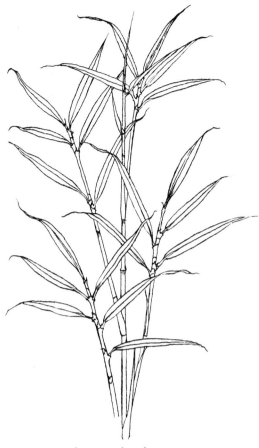

Japanese bamboo,
Pseudosasa (Bambusa) japonica

109

Joseph Hooker in the Himalayas, from a painting by Frank Stone

Dried specimens of
Rhododendron roylei
brought back from
Sikkim in 1848 by
Hooker; now in the
Herbarium at Kew
Gardens

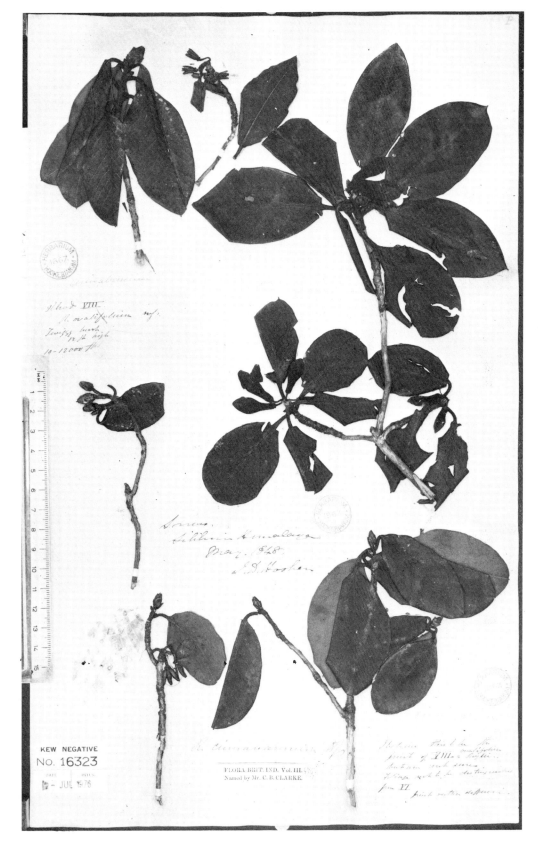

Above: Charles Maries. *Above right:* Robert Fortune. *Far right:* the frontispiece to Robert Fortune's book *A Journey to the Tea-Countries of China*, published in 1852

Tree paeony, *Paeonia suffruticosa*

Robert Fortune, a pioneer plant-hunter in China, also worked in Japan in the 1860s, and despite restrictions on his movements was even more successful than von Siebold and Veitch. Then, for a decade the struggle between the Mikado and the Shogunate, with local war-lords on the rampage, the murder of an English visitor and the reprisal – the bombardment of Kogoshima by the Royal Navy – made Japan unsafe for plant-hunters. But in 1875 work was resumed there by Charles Maries, for Veitch & Co., who also collected in China. Yet more trees and shrubs and lilies from the temperate Far East began to make a place for themselves in English gardens, parks and woodlands.

This work of finding new plants which would flourish in Britain and sending them home was far from being a simple matter of walking the countryside, choosing suitable plants and packing them off to Europe. The plant-hunters worked in rough, wild country; they had to have native helpers to act as porters and guides. They frequently had to face hostile populations, bandits, pirates and rapacious local war-lords. They suffered from tropical heat, arctic cold, hunger and thirst and fever and the attacks of wild beasts. It might be relatively easy to send home seeds and bulbs; but in every case there had to be a careful description of the plants. As they were also expected to collect herbarium material the delicate task of pressing flowers and leaves had to be done in very difficult conditions. In cases where only live plants would do there was the problem of keeping them alive during the remainder of the expedition and then during the long voyage home by sea, first in slow sailing ships and later in steamers which were not, in their early days, much faster. Before the invention of the Wardian case a majority of plants died; after that invention the survival rate was high but the cases were heavy, awkward and fragile objects to carry, on mule-back or porter-back, over torrential rivers and rugged mountain ranges.

The work of plant-collecting in China had, until the infamous Opium War of 1840 in which our military power was used to force the Chinese government to accept opium imports from India, been as much ham-

112

Carl Maximowicz

pered by restrictions on travelling as it had been in Japan. But some relaxation followed the treaty imposed on China – it included the cession of Hong Kong – and the Royal Horticultural Society sent Robert Fortune, superintendent of their Chiswick hot-houses, to see what he could do. He spent nearly twenty years collecting in China and Japan, sending home mahonias, tree paeonies, weigelas, forsythia, the winter-flowering jasmine and hundreds of herbaceous species. Among his dangerous adventures was a fight with Chinese pirates. He was responsible for the establishment of the tea industry in India.

The first systematic collecting in Northern China was done by the Russian botanist Carl Maximowicz whose first Chinese journey was made in 1853. Following the crushing of the Taiping rebellion by General Gordon, in the imperial service, penetration of the interior became possible for European botanists and collectors. In 1866 Father Armand David, science master at the French school run by the Lazarist mission in Peking, made the first of his three journeys of plant exploration in Inner Mongolia, in Shensi and on the frontier with Tibet. His discoveries led Veitch to send E. H. Wilson to China in 1899, specifically to collect seeds of *Davidia involucrata*, the handkerchief tree, first grown from seeds collected by David and sent to Maurice de Vilmorin, the greatest patron of plant-hunters and plantsmanship in France. Wilson went to Yunnan, where he stayed with Augustine Henry, the great Irish plantsman whose knowledge of the country and whose plant discoveries were of the greatest service to several collectors working in China. From there he went to Ichang on the Yangtze. He found **davidias**; and returned to England with thirty-five Wardian cases and seeds of over three hundred new species, as well as a vast load of herbarium material.

Maurice de Vilmorin

Augustine Henry

Father Armand David

8432

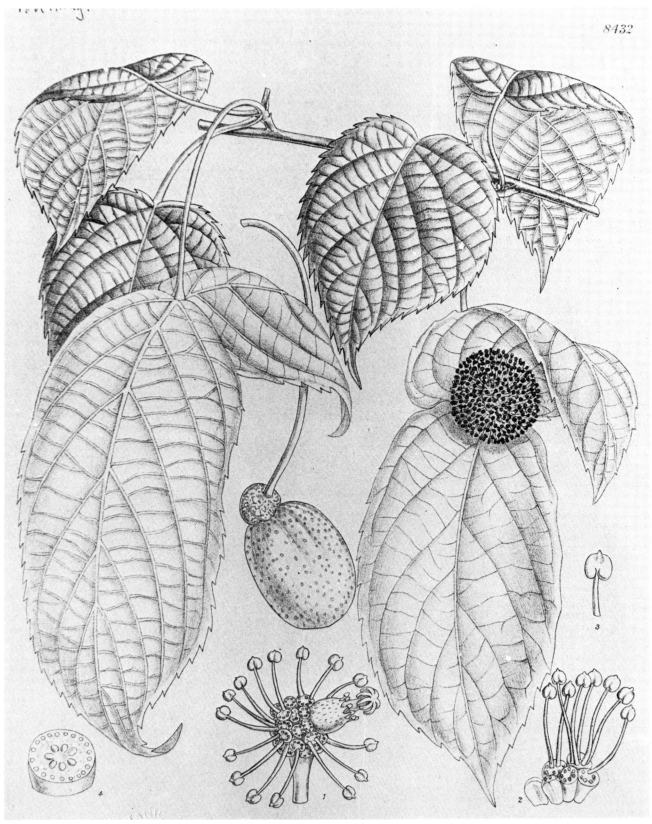

Handkerchief tree, *Davidia involucrata* 'Vilmoriniana'

115

Above: detail of the handkerchief tree.
Right: E. H. Wilson and, *below:* Professor
Charles Sargent, first director of the
Arnold Arboretum at Harvard University

He made two more equally fruitful expeditions for Veitch and then went to work for the Arnold Arboretum in Boston, Massachusetts, for which he was again successful. Like so many plant collectors he endured hair-raising adventures and hardships; they were a price he stoically paid for the enriching of the cultivated florae of Europe and America. Professor Sargent's *Plantae Wilsoniae* (Arnold Arboretum), where the plants Wilson brought back are listed and described, fills three large volumes.

The Dutch-American Frank Meyer, collecting new food and fibre plants for the U.S. Department of Agriculture in Russian Turkistan, China, Korea and Manchuria, also sent home a large number of new ornamental plants and trees, which soon reached England from America. Meyer was found drowned in the Yangtze Kiang in 1919; how it happened has never been discovered. The U.S. Department of Agriculture sent another collector to the East, the Austrian-born Joseph F. Rock. His specific mission was to find the *Kalan* tree, source of chaulmoogra oil, which Indians believed was a cure for leprosy. Beginning in 1920 and collecting on the way, he travelled in Siam, Yunnan, Szechuan and Muli on the Tibetan frontier, sending a stream of new plant material back to America and Europe.

In contrast to such professionals, England's outstanding collectors at

Left: Joseph F. Rock. *Above:* Frank N. Meyer in Shansi, China. *Below:* Reginald Farrer

Potentilla fruticosa

this time were amateurs. Reginald Farrer went collecting in Kansu in 1914–15. He was the only plant-collector until Francis Kingdon Ward, to write a good book about his travels, bringing home new buddleias, viburnums and potentillas. E. H. M. Cox collected in Upper Burma in 1919, bringing home barberries, glorious new rhododendrons, a splendid *Magnolia* and the magnificent *Juniperus coxii*. A year later Farrer, working the same territory, died there of a throat infection. His greatest contribution was to rock-gardening, which he had practised since his fourteenth year. H. J. Elwes was a very rich man who began life as a Guards officer, sent in his papers because soldiering bored him, and went hunting plants and animals in China and Tibet, Japan, Siberia and Russia, India, Turkey and both Americas, meanwhile teaching himself botany and zoology. His *Monograph of the Genus Lilium* is his magnificent monument as a botanist.

George Forrest, a Scot born in 1872, made seven expeditions and worked for a total of twenty-eight years in the Far East, chiefly in Yunnan. He introduced several hundred new species of rhododendron, many more than any other collector. He was hunted by bands of fanatical Tibetan monks – his two companions, French missionaries, were caught, tortured and murdered, along with sixty-eight other people, when escaping from the French mission at Tseku. He escaped, only to die of a heart attack in Teng-huey in 1932.

Right: George Forrest and, *far right, Rhododendron bullatum,* one of the many new species of rhododendron discovered by Forrest

Frank Kingdon Ward

Himalayan blue poppy,
Meconopsis betonicifolia

Forrest's greatest successor was Frank Kingdon Ward, born in Manchester in 1885. His plant collecting was done in China, Upper Burma and Tibet. He was as considerable an explorer and writer as he was a plant-hunter. We owe him, among other things, the blue 'poppy', *Meconopsis betonicifolia*.

China was, without question, the land of election of the plant-hunters who were collecting the material to enrich our cultivated – and sometimes wild – flora in the early years of this century. But it was by no means the only one. In 1897 Harold Comber was born, son of the head gardener of Nymans, the great Sussex garden created by Colonel Messel. He was trained as a gardener and botanist (Edinburgh Royal Botanic Garden). In 1925 Lord Aberconway – creator of the Bodnant Gardens in North Wales – organized a syndicate to send Comber to look for plants in the Andes; and on a second expedition he went to Tasmania. Many of his trees and shrubs have proved hardier in England than they were expected to be. And so have some of the rhododendrons, viburnums and cotoneasters introduced from the mountains of northern India by R. E. Cooper, who was plant-hunting there between 1913 and 1915.

* * *

The enrichment of our natural flora began, then, so long ago that we shall never know the names of those who started the work. Although, as we shall see, a great number of their introductions have become more or less naturalized here, most of them remain confined to gardens, parks and tended woodlands.

Strictly speaking the word flora in our context should be applied only to the plants which grow wild in the country in question, so that I have repeatedly used the words 'cultivated flora' to cover myself. But this book is about the change in the visual aspect of England wrought by the introduction of hundreds of plants now familiar elements of the scene. In a sense, the 'flora' of a land as tamed, made over by man and urbanized as England, is composed of the plants we are so used to seeing that they are 'natural' to the scene.

Here I should like to make a comparison with a part of the world which is much 'older' than England, in the sense of having been tamed, made over by man and urbanized many centuries earlier: the northern Mediterranean basin. Ask anybody familiar with it to name the elements of the French–Italian–Spanish Mediterranean scene, of the flora as we are aware of it in the visual aspect of the country casually taken in as they pass: they will include the olive, the Italian cypress, the prickly pear, the agave, the fig and the vine. Without those six, the Mediterranean littoral is 'inconceivable' – it would, as it were, be somewhere else.

Well, taking these six elements of the visual aspect of those lands in chronological order of introduction: the fig was a plant of the Tertiary epoch; it was wiped out by the last Ice Age except in the Canary Islands whence, spreading to north-west Africa, it slowly re-colonized the southern shores of the Mediterranean and was introduced to the northern shores, spreading steadily eastward. The olive and the vine, both west Asian natives, were introduced to the western Mediterranean

Harold Comber

scene not earlier than 700 B.C. and not later than 400 B.C. by the Greek colonists of Sicily and Calabria. With them may have come the 'Italian' cypress – a native of the mountains west of the Caspian Sea in Iran – but at a guess it came later, because whereas olive and vine are food-and-drink plants, the cypress is an 'industrial' plant, introduced for its value in ship-building – it provides excellent masts. As for the prickly pear (*Opuntia*) and the agave, since they are both middle American natives they cannot possibly have been introduced before about 1550, and probably much later.

In short, the 'typical' plants of the south of France and Italy are no more natural to the scene than evergreen oaks or Scots pines in England. And in the long run it is not of material importance whether a species of plant is newly introduced into a landscape by a fruit-eating or nut-eating bird or by man.

There is, however, an obvious difference between plants 'naturalized' into our flora some time following their introduction and those which survive in it only because they are cared for. For the reason already given we are not much concerned with that difference, but it is certainly important in the context of any discussion of what we mean by native 'flora'.

121

Chapter Eight

Summing up

WHAT does all this amount to, supposing we are required to sum it up in a brief statement outlining the history of the flora of England?

Our present wild flora contains about 2000 species of trees, shrubs, herbaceous perennials and annuals and our cultivated flora is without doubt greatly in excess of this number.

Before the onset of the Ice Age about a million years ago, or in one of the warmer interglacials, and after the tropical plants had been driven from England by steadily declining mean annual temperatures, some hundreds of species belonging to familiar generic types were already in being and growing in England. Although fossils of only about 500 of such more or less familiar species have been found and identified, it may well be that there were more than that.

During the Ice Age advance of the glaciers which preceded the interglacial in which we are living now the great majority of those species were exterminated in England, but persisted in the lowlands of south Europe, North Africa, West Africa and the Atlantic islands to which they had retreated because the cold was less severe and instead of ice there were heavy and persistent rains.

As the ice withdrew from England and the climate became slowly more genial, say about ten thousand years ago, the hardy pine trees, spreading northward from regions where the warming up of the climate came earlier, invaded England and spread until they covered it with coniferous forest of the kind which can still be seen in north Europe. Then, as the climate became still warmer, while the pines withdrew northwards, the deciduous trees and hundreds of species of lesser plants associated with them in plant communities moved in to the south of England; some had been here before the last Ice Age, others were genuine newcomers. In due course England was covered with mixed deciduous forest excepting on the chalk uplands, with oak predominating.

At a time four or five thousand years ago, when man began to make his first impression on the flora, whether deliberately or as a consequence of his activities, the native flora was probably composed of something like 1500 species of temperate-zone trees, shrubs and herbaceous plants, and, of course, hundreds of species of cryptogams: it was much less numerous than the flora of continental Europe, partly because before the migration of plants was complete the Channel had become wide enough to interpose a barrier to plant immigration.

In the course of the next several thousand years, as a result of sea-borne accidental imports of seeds in trading and war ships, but also as a result of man's deliberate action, about five hundred species were added to the wild flora. And during the last twenty centuries about two thousand more species were added to the whole flora (that is to the wild and cultivated flora considered as a whole), by the deliberate introduction of plants for food, animal fodder, fibres, oil-seeds, timber and garden ornamentation.

If, then, we consider our flora as a whole and as it now is, we can say this: that the proportion of that given to us by nature to that given to us by our own activities is roughly three to five.

But if we consider only the wild flora, then we owe three-quarters to nature and a quarter to ourselves.

Acknowledgements

*The colour photographs are by Eric Crichton
except where otherwise noted.*

The publishers wish to thank the following for permission to reproduce black and white illustrative material: Academisch Historisch Museum, Leyden, for page 106 *below left* and *below right*; The Arnold Arboretum of Harvard University for page 116 *all three*; Ashmolean Museum, Oxford, for page 98 *below left* and *below right*; British Library Board for page 96 *below*; page 110 is by permission of the Trustees of the British Museum; pages 100, 102 *below right*, 103 *below* and 109 *below right* are by permission of the Trustees of the British Museum (Natural History); page 104 *below* is from *The Golden Age of Plant Hunters* by Kenneth Lemmon, Courtesy of J. M. Dent; The Forestry Commission for page 107 *below left*; Greater London Council (Print Collection) for page 98 *above*; Historical Society of Pennsylvania for page 101 *left*; Hunt Institute for Botanical Documentation, Carnegie - Mellon University, Pittsburgh, Pennsylvania, for page 117 *above right*; pages 104 *above* and 107 *below right* are by permission of the President and the Council of the Linnean Society of London; The Mansell Collection for page 99 *above right* (Gladstone Conservatory, Liverpool); Museum National d'Histoire Naturelle for page 114 *below left*; National Botanic Gardens, Dublin for page 114 *below centre*; The National Portrait Gallery, London, for pages 99 *above left* and *centre*, and 102 *below left*; Radio Times Hulton Picture Library for pages 99 *below left* and *right*, 101 *above right*, 102 *above left*, 103 *above*, 106 *above right*, 107 *above* and 108 *above left* and *right*; Royal Botanic Gardens, Kew for pages 106 *above left*, 108 *below left*, 109 *below left*, 111, 112 *above left and right*, 113, 114 *above and right*, 115, 117 *below right*, 120 *above* and 121; Royal Library of Stockholm for page 105 *above left* and *right*; John Topham Picture Library for pages 62 *below*, 63 *above left* and 71 *below left*; The United States Department of Agriculture for page 117 *left*.

The publishers also wish to thank the following for permission to reproduce additional colour material: John Beecham for page 47 *below left*; Bruce Coleman Ltd for pages 71 *below right* and 82 *above left*; C. M. Dixon for page 95 *above*; Anthony Huxley for pages 47 *above left*, *above right* and *centre*, and 75 *below*; Frank W. Lane for page 47 *below centre* (Joan Hutchings Photo); Scala Istituto Fotografico Editoriale for page 95 *below*; Harry Smith Horticultural Photographic Collection for pages 70 *centre* and 82 *above right*.

Josephine Rankin based her paintings on live specimens and on illustrations in the following:

Wilfred Blunt, *The Art of Botanical Illustration*, Collins, 1950.
Gerta Calmann, *Ehret: Flower Painter Extraordinary*, Phaidon, 1977.
Brenda Colin, *Trees for Town and Country*, Lund Humphries, 2nd ed., 1950.
Herbert L. Edlin, *Know Your Broadleaves*, HMSO, 1968; *Know Your Conifers*, HMSO, 1970.
Roy Hay and Patrick Synge, *Dictionary of Garden Plants*, Michael Joseph, 1969.
F. G. Heath, *The Fern World*, Sampson Low, 1877.
S. Hibberd, *The Fern Garden*, Groombridge & Sons, 1869.

Gertrude Jekyll, *Wood and Garden*, Longman, 1899; *Home and Garden*, Longman, 1901; *Roses for English Gardens*, Country Life, 1905; *Children and Gardens*, Country Life, 1905; *Wall and Water Gardens*, Country Life, 1905.

R. W. C. A. Jones, *Flowers of the Field*, Routledge, 1911.

W. Keble-Martin, *The Concise British Flora in Colour*, Michael Joseph, 1965.

S. Millar-Gault and P. M. Synge, *Dictionary of Roses*, Michael Joseph, 1971.

S. Millar-Gault, *Dictionary of Shrubs*, Michael Joseph, 1976.

B. E. Nicholson, *The Oxford Book of Food Plants*, Oxford University Press, 1970.

George Nicholson, ed., *Illustrated Encyclopaedia of Gardening* (12 vols.), L. Upcott Gill, 1885.

C. P. Petch and E. S. Swann, *Flora of Norfolk*, Jarrold and Sons, Norwich, 1968.

Oleg Polunin, *Flowers of Europe, a Field Guide*, Oxford University Press, 1969.

Oleg Polunin and Anthony Huxley, *Flowers of the Mediterranean*, Chatto and Windus, 1974.

W. Robinson, *The English Flower Garden*, John Murray, 1889.

Royal Horticultural Society, *The Garden* (monthly).

Edward Step, *Wayside and Woodland Ferns*, Frederick Warne, 1908.

Edward Step, ed., *Trees and Flowers of the Countryside*, (2 vols.), Hutchinson.

G. Stuart Thomas, *Climbing Roses Old and New*, Phoenix House, 1965; *The Old Shrub Roses*, Phoenix House, 4th ed., 1963.

Index

General Index

Figures in *italic* type indicate the inclusion of an illustration

Index of Scientific Names

Figures in *italic* type indicate the inclusion of an illustration